Having gained his doctorate at Cambridge, **Roy Porter** was University Lecturer in History there before moving in 1979 to The Wellcome Institute for the History of Medicine. He has co-edited the *Dictionary of the History of Science* (1981); his own books include *English Society in the Eighteenth Century* (1982) and *A Social History of Madness* (1987).

Roy Porter's elegant study of Edward Gibbon is the first of its kind in almost twenty years. Neither a full-length biography nor a specialist monograph, it is a study of Gibbon as historian: a product of his own time, and an enduring voice in our own. His *History of the Decline and Fall of the Roman Empire* is without doubt the most distinguished and best-known work of history in the English language; but how did Gibbon come to write it? Roy Porter explores the magical connotations of Rome and its Empire for the civilization of the Enlightenment, and gives an intriguing account of Gibbon's own very odd childhood and adolescence which turned him into a solitary scholar, with deeply-held views about religion and political power. Rome, Gibbon eventually decided, would be the best challenge to his powers and his best hope of fame. Roy Porter's incisive portrait examines the special – and controversial – qualities of Gibbon as historian, showing the man, the mind and the history as inevitably, complexly, intertwined.

Edward Gibbon

Roy Porter

Edward Gibbon : Making History

Weidenfeld and Nicolson
London

First published in Great Britain in 1988 by
George Weidenfeld & Nicolson Limited,
91 Clapham High Street, London SW4 7TA

Copyright © 1988 by Roy Porter

Printed in Great Britain at The Bath Press, Avon

Contents

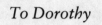

To Dorothy

Preface

How did Edward Gibbon become a historian? What was his idea of the past? Why did he write the kind of history he did? This small book has big questions to discuss. It is an intellectual biography, and a selective one at that, probing into only one aspect of the mind of a man who also had much to say about people, politics and writing. It is not a biography as such, though I, like many others, have heard the siren luring me to try my hand at reconstructing 'the Gib', Byron's 'lord of irony', the autobiographer who revealed yet concealed so much. Fortunately, Gibbon's life story has already been told in a string of admirable biographies; and the older classics, such as those by Low and Young, have now been joined by more recent rivals, benefiting from new research into unpublished papers (in the British Library and elsewhere), and, perhaps, from the insights of modern psychiatry.

Gibbon scholarship has been first-rate, and anyone familiar with it will instantly see how deeply indebted I am to the investigations and insights of a cohort of Gibbon experts. The format of this series precludes me from acknowledging my debts, and occasional dissent, point by point in conventional footnotes; but I trust that the Bibliographical Essay which rounds off this sketch will make those debts quite clear, as well as offering signposts for readers wishing to follow up particular themes in greater detail than space permits. I believe that one dimension of Gibbon's life and work has been unduly neglected: understanding the ways in which his career as historian and his idea of history-writing drew upon, yet diverged from, those of other contemporary British and Continental historians. Gibbon's biographers have treated him rather in isolation from his times. I have attempted to remedy this omission somewhat by juxtaposing Gibbon against the common practice of history and the common sense of the past in early modern England. By doing so, I hope to dispel a certain anachronism which often creeps into Gibbon scholarship. For his powers as a historian are frequently judged – and judged unfavourably – against nineteenth- and twentieth-century methodologies and philosophies of history. Rather than trying to measure him thus in terms of the evolution of today's norms, I have instead tried to probe the problems and principles of history as they appeared in Gibbon's own time.

As a scholar, Gibbon was a solitary. We dwarfs standing on giants' shoulders can write history thanks only to the unstinting stimulus, expertise, generosity and encouragement of friends and colleagues. Logie Barrow, Janet Browne, W. F. Bynum, Estelle Cohen, Mark Goldie, Margaret Kinnell, Sue Limb, Gertrude Prescott Nuding, Clarissa Campbell Orr, G. S. Rousseau, Christine Stevenson, Sylvana Tomaselli, Jane Walsh, and Andrew Wear have all read this manuscript and have offered extremely fruitful criticisms. I am most grateful for their help, and hope I have learned from their advice; remaining errors are of course my own. As always, the Wellcome Institute for the History of Medicine has provided the perfect environment for research and writing, and Jean Runciman has composed an excellent index. As always, Juliet Gardiner has proved the perfect editor.

Note: Perhaps partly because he was bilingual in French and English, Gibbon's spelling was often eccentric. In quoting him, I have retained his spellings.

Many people are present. They are all walking to the left round a square. The dreamer is not in the centre but to one side. They say that a gibbon is to be reconstructed.

From one of Carl Jung's dreams

Introduction

When John Perceval, son of the assassinated Prime Minister Spencer Perceval, went mad in the early 1830s, he was detained in Brislington House lunatic asylum. Restrained and restless, he rapidly grew bored. He wanted to read, to exercise his mind. One book in particular he requested: Edward Gibbon's *Decline and Fall of the Roman Empire*.

At roughly the same time, a young artisan, Joseph Gutteridge, was trying to make his way in the world. An autodidact, Gutteridge was dedicated to strenuous self-improvement. 'During the early years of married life,' he later reminisced in his autobiography,

> I gave myself up to extensive reading upon all subjects that came within reach, but my greatest pleasure was in ancient history and the mythologies of past ages. Favourite books were Gibbon's 'Decline and Fall of the Roman Empire', Volney's 'Ruins of Empires' and 'Laws of Nature', Rollin's 'Ancient History', and Mosheim's 'Ecclesiastical History'. With these I was highly delighted, and a great part of the matter contained in them was committed to memory.

Gutteridge's absorption in Gibbon is easily explained: self-made men liked seeing the powers of evil exposed and the mighty fallen. Perceval's yen to read him is more puzzling: as a fervent Evangelical he surely abominated Gibbon's sneers against the faith; perhaps he simply wished to feast his intellect upon the greatest prose epic in the English tongue. What is quite clear, however, from the reactions of these two readers, and from scores of others, is that the *Decline and Fall* — six volumes, a million and a half words, and eight thousand footnotes — had already secured a niche in the historical pantheon comparable to Shakespeare in drama and Milton in poetry. Gladstone ranked Gibbon one of the three greatest historians of all time.

Gibbon did not think he was born great. He did, however, come to feel that history was his destiny: '*I know* by experience that from my early youth, I aspired to the character of an historian'; and he set about achieving greatness – as early as 1773, three years before the *Decline and Fall* began to be published, he was referring to it as 'my great work'. Nor was he disappointed. 'Lo! A truly classic work has appeared,' exclaimed the astonished Horace Walpole on reading the first volume in 1776 – high praise indeed from a critic so often catty about fellow authors. And the public echoed Walpole's enthusiasm. The publisher Cadell boasted that the book 'sold like a three penny pamphlet on current affairs' and that 'within a fortnight not a single copy remained' of the original print-run of a thousand. 'The first impression was exhausted in a few days,' glowed Gibbon in his *Memoirs of My Life*,

> a second and third Edition were scarcely adequate to the demand; and the bookseller's property was twice invaded by the pyrates of Dublin. My book was on every table, and almost on every toilette; the historian was crowned by the taste or fashion of the day.

Understandably, the following five volumes – he was 'brought to bed of twins' in 1781, and delivered of the last three in 1788 – did not make quite the same splash. 'Another damned thick square book!' was the Duke of Gloucester's crusty reaction to Gibbon's fertility. But by the time of his death in 1794 – by then abridgements had already started to appear – Gibbon's place was securely established in the triumvirate of British historians, alongside the Scots William Robertson and David Hume; he was recognized as incomparably the greatest English historian. In fact he had become, as he has been known ever since, '*the* historian', and his very title entered the national idiom and heritage: people who cannot name any other work of history can rattle off the phrase *Decline and Fall*. Moreover it has held its reputation amongst many scholars too: Gibbon remains 'the greatest of the historians of the Enlightenment,' argues Hugh Trevor-Roper, 'the only one of them who is still read not only as a stylist but also as a historian'.

But is this compliment literally true? Here Geoffrey Elton contradicts: 'hardly anyone reads [him] any longer' – a fact regarded by Elton as especially damning, since in his eyes 'readability' must be Gibbon's main recommendation, his scholarship having now been superseded by the labours of two centuries. Has Gibbon then been overrated? Have we at long last witnessed *his* decline and fall?

Indeed, despite the applause of 'men of letters, men of the world, and even ... fine feathered Ladies', there has always been a counter-current of doubt and damnation. Right from the beginning, Gibbon's sneering and cynical interpretation of the triumph of Christianity drew fire from ecclesiastical critics; most of these, however – Prebendary Travis, Davis, Apthorp, Dr Randolph and 'poor Chelsum' – were nonentities who merely let off blanks. But weightier figures were also to register their disapproval. Coleridge dismissed Gibbon's prose style as meretricious: 'when I read a Chapter in Gibbon I seem to be looking at a luminous haze, or fog' (Coleridge would know). More severely, he accused Gibbon of stooping to historical scene-painting ('all larger than life, or distorted and dis-coloured ... all is scenical') to hide a want of analytical profundity, a failure to grasp the grand forces of destiny:

> To call it a History of the Decline and Fall of the Roman Empire! Was there ever a greater misnomer? I protest I do not remember a single philosophical attempt made throughout the work to fathom the ultimate causes of the decline or fall of that empire.

Coleridge's indictment – that the shallow Gibbon had 'no philo-sophy' – has hardened into a kind of orthodoxy. It is conceded that Gibbon was of course a grand master of narrative, unequalled in dramatic flair and sardonic pen-portraits. But was he able to probe beneath the surface of events and grasp the true causes of historical change? Many modern scholars think not. Martine Watson Brownley has condemned his 'failure to surmount fact and to move towards significant generalization', explaining that this 'limitation' followed from his not being 'an abstract and speculative thinker' ('his lack of trust in abstract thinking seriously limited his philosophical depth'). Some organic or synthetic power of mind was, she believes, wanting: 'He lacks the cohesive imagination and the ability to rise above historical detail which are necessary to derive significant generalizations from the specific.' Brownley's verdict is that Gibbon combined 'factual strength and philosophical weakness': as a result of the latter, 'he does not often enough sur-mount facts to produce coherent and satisfying historical overviews.' A similar judgment on his supposed lack of analytical depth has been passed by Joseph Swain, who has noted that:

readers of Gibbon's *History of the Decline and Fall of the Roman Empire* will be disappointed if they expect to find in it a closely reasoned philosophical or sociological treatise on the causes of Rome's collapse. Nowhere in his six volumes does he make a full and explicit statement of what, in his opinion, were the great determining causes of that catastrophe. He does not even discuss such elementary but fundamental questions as the meaning of the word 'fall', and exactly what fell.

The fundamental question of why Rome fell, he concludes, is glossed over in some 'sketchy remarks' in a 'few paragraphs of the airy "General Observations"' which round off the third volume; and numerous other modern commentators join Swain in finding Gibbon's analysis there hackneyed and superficial: 'the chapter abounds in pompous inanities, many of which are almost wholly beside the point,' judges Stephen Graubard.

Altogether, when evaluating Gibbon's techniques, scholars emphasize that history-writing has moved on. His explanatory parameters were too narrow. When addressing the decline of Rome, he was far too content (argues Glen Bowersock) to parrot Tacitus' rhetoric of 'corruption'; instead of indulging in such moralizing, Gibbon should have explored the socio-economic causes of imperial decline more rigorously. Along similar lines Per Fuglum contends that Gibbon finally failed to make sense of his subject because he neglected the role of the middle class in history and said little about the 'people': overall 'Gibbon saw no meaning in history.' 'Gibbon was content,' argued J. B. Black, 'to set the puppets in motion; he left it to succeeding writers to penetrate behind the scenes and discover the laws which determined the course of events.'

It has furthermore been claimed that Gibbon's prejudices fatally flawed his historical understanding. The lover of Rome never disguised his contempt for the superstition and fanaticism of the Christian Churches, for the wasted centuries of Byzantium ('a tedious and uniform tale of weakness and misery'), and for the 'gothic ages'. Majestically Olympian, Gibbon 'too assiduously assumed the task of a moral judge', thought Fuglum, whereas (scholars agree today) the historian's job is not to judge but primarily to understand. As a result, Christopher Dawson contended, when faced with the rise of Christendom Gibbon, blinkered by basic 'limitations of spiritual vision and historical imagination', fails to 'understand the past' and 'only succeeds in explaining it away': 'To Gibbon the story of the Christian Empire and the civilization to which it gave birth is nothing but the history of an illusion.' Overall, suggested Herbert Butterfield, Gibbon suffered from the weakness of the 'Whig interpretation of history', judging the past by the standards of the present.

Thus Gibbon, critics allege, lacked both analytical power and empathetic vision. He also lacked, it is said, that most vital tool in the modern scholar's kitbag: proper methodological rigour. By this it is not meant that he was a slapdash scribbler who got his dates wrong and misquoted his authorities. These were indeed the charges levelled against him by his clerical enemies, but falsely so; both Porson and Robertson, two of the most exacting contemporary scholars, did Gibbon simple justice when they complimented him on his exceptional scrupulosity with his sources. Rather it is argued that Gibbon's technique was flawed because he made shift with an approach to research which was becoming old-fashioned even in his own day – Elton has called it history's 'prehistoric age' – a research method which the scholarly advances of the next generations rendered obsolete.

Gibbon chose to work solely from printed materials. He felt happiest with literary sources, above all contemporary histories from Livy and Tacitus through to the medieval chroniclers. Addressing these, he did not perceive the need for systematic, scientific source criticism (*Quellenkritik*) as understood by the next century. He ignored manuscripts; he never fully mastered the new auxiliary sciences of palaeography and diplomatics; and though he drew upon the findings of scholars who had pioneered the deciphering of non-literary evidence – coins and medals, architecture and inscriptions – he made no discoveries of his own in these fields. In other words, Gibbon did not add to our factual knowledge of the past.

Indeed, he rarely stirred beyond his own private library. He beavered away in splendid isolation, and apparently never discussed his work with fellow scholars while it was in progress. An early decision not to learn German – it was far too barbarous a language – insulated him from some of the bracing scholarly breezes blowing on the Continent; indeed, he was unable to read the most perceptively critical reviews of his own masterpiece.

Had Gibbon remedied these 'defects' in his technique, would he have written a different history? It is doubtful. In any case, as hardly needs saying, the marvel is how much Gibbon read, knew and understood, not the fact that certain chinks remained in his scholarly armour. What is beyond dispute, however, is that historical research was transformed in the generations after Gibbon. Archives were stormed and ecclesiastical libraries ransacked for manuscripts; critical editions were prepared deploying advanced techniques of source criticism, drawing upon the disciplines of philology, palaeography and so forth which had still been in their infancy in Gibbon's day. A century after Gibbon's death, J.B.Bury prepared a revised edition of the *Decline and Fall* in which, from the standpoint of positive historical knowledge, he could demonstrate point by point how a century of collective professional activity had chipped away many of the details of Gibbon's edifice. That century of work had, however, confirmed still more, and left the broad architecture astonishingly intact.

Gibbon had no followers and founded no school. Small wonder perhaps that today's academic historians find him difficult to place in the development of the discipline of history. For Elton, doyen of today's historical professionalism, the true masters of the historian's craft in Britain were Frederick Maitland and Lewis Namier, pioneers of painstaking original research in depth. As we have seen, Elton thinks Gibbon's main recommendation today is his style – a backhanded compliment, often paid. Thus, according to Cochrane, 'the permanent and essential value of his work is as literature.'

We have long been haunted by the ghost of the Roman Empire; the *Decline and Fall* has been called by Carl Becker more than just a tale of that empire, but rather 'a memorial oration . . . commemorating the death of ancient civilization'. I do not intend this book to be such an oration over Gibbon, either celebrating him as the father of English historiography, or elegizing him as the 'last of the Romans'. Nor shall I try the very unGibbonian feat of raising the dead, by contending that he was a wiser historian than his more expert supersessors – though it would be hard to name another author who has since made a better job of writing the history of fifteen hundred years of Mediterranean civilization: the career of the late Fernand Braudel offers food for thought. Least of all do I want to insult Gibbon by suggesting that he should be read principally for his figures of speech.

There is, however, a very compelling reason why all who care about history should read the *Decline and Fall* and explore Gibbon the man. It is because his life and work afford us an extraordinarily

6

vivid, well-documented and often poignant example of a powerful mind making history; they reveal how a massive intelligence, steeped in the culture of the present, illuminates the dim and difficult data of the past, through exercising 'the sportive play of fancy and learning', weaving them into a fabric which joins the vision of myth with the logic of science. We see culture, ideology and erudition coalescing in a way fundamental to the historical process.

For well over a century now, the alluring idea of establishing history as a science (Bury himself argued that it is 'no less, no more') has had many advocates. The claim is clearly tendentious; yet few would deny that professionalization, and such new sub-disciplines as cliometrics, have in some ways rendered the production of historical knowledge more objective. Yet at the same time we have become more sensitive to the ideological messages of our pictures of the past, their use and abuse. We are thus faced with a profound paradox: we pursue objectivity to exclude 'bias', yet in practice that objectivity may not eliminate 'bias' but actually write it in.

Thus the problem arises of how a subjective author can produce an objective ('authoritative') creation in history. It is a paradox of which we are aware, but which we rarely have the opportunity of studying in the flesh. Perhaps overburdened with consciousness, few historians write their autobiographies; few today are commemorated after their death with biographies; and most do not leave us their research notes or their annotated libraries. We are in the unsatisfactory position of eagerly subscribing to E.H.Carr's dictum (to understand the history, first study the historian), without actually having many studies of historians to hand.

Gibbon offers a splendid exception. For the scythe of time has left us with a tempting, if dangerous, simplicity: we have a manageable sheaf of letters and papers (most now edited according to those exacting standards which the last century championed). In Gibbon's case, we possess, not as with most historians today dozens of scattered research articles, but a single *magnum opus*, buttressed by a few other writings which can be seen as apprentice works. If ever there was a classic 'one-book man', someone whose fame, even his very identity, is utterly associated with one great feat of vision and craftsmanship, it surely must be Gibbon. Indeed, the *Decline and Fall* can be read as a form of running commentary on the emergence of the historian's powers, on the forging of his identity as a historian, indeed as 'the historian of the Roman Empire'. All great books may be implicitly autobiographical; but the *Decline and Fall* is perhaps more so, and more self-consciously so, than most. His personality, prejudices and preoccupations are exposed throughout his text, and pepper his footnotes, the bon viveur's table-talk.

Gibbon kept copious, introspective journals for some of the most decisive spells of his life; and then, living in Lausanne after the completion of his masterpiece, decided to write what became a proliferation of autobiographies. No fewer than six substantially different versions were penned; who knows how many more reworkings would have been recorded had he not been cut off in his prime in 1794, aged fifty-six? What these show – individually and collectively, explicitly and between the lines – is that Gibbon was extraordinarily self-aware as to how history was necessarily the creation of the shaping intellect, indeed of the imagination; it was the fabrication of meaningful memory every bit as much as it was the recording of *res gestae* (things done). Great actors performing on the world's stage made history. But only thanks to historians. The crucial distinction which had to be preserved by the historian was between making history and making it up.

Gibbon's autobiographical writings, as I shall stress in Chapter 2, are concerned to trace that mysterious marriage of necessity and circumstance through which he became 'the historian of the Roman Empire'; but they are no less preoccupied with the meta-problem lurking behind that: that of his own varying, retrospective *perceptions* of how he became that historian. Each stab at writing his *Memoirs* perhaps released more associations, and changed what he 'remembered'. Gibbon autobiographizing 'alone in Paradise' in the autumn of his days calls to mind Proust *à la recherche du temps perdu*, but also, more ludicrously, his fictitious contemporary, Tristram Shandy, for whom writing his own life swamped living it. To Gibbon, as he looked back, all could seem so assured, so predestined: he remembered that from his youth he knew that he was to be a historian; he remembered, or fancied he remembered, or fantasized – want of evidence leaves the matter irresolvable – how one particular evening, 15 October 1764, as he sat 'amidst the ruins of the Capitol', the idea of writing the great work first sprang into his mind, twelve years before a word of it appeared. But all of that could also seem so accidental, a trick of fate (could it even be Providence?). His recollection of his early life was that he had been the sport of fortune in a way which echoes the young

8

Tristram Shandy, reminding us that Gibbon thought that life often imitated art. Both escaped childhood barely alive and hardly educated. Tristram lost his name and his nose and his foreskin; the young Edward of the *Memoirs* also nearly lost his name, lost his breast and milk, lost his mother to his father, lost his brothers and sister (all six had died), lost his home, his education, his health, his university, his religion (more than once), his country and, soon after, his sweetheart. And yet, by a flick of historical interpretation, all these losses could be made into gains, emancipations even; he had lost his chains and won his freedom, lost a past and won a future.

Gibbon thus probed the problem of creator and creation as he wrote the history of the self who had written the history of Rome. Ever self-aware, but self-perceptive with an artistry which dares to draw attention to his own blind spots, he pondered how his mind had come to write that book, the history of his own mind. In an early unpublished manuscript, the 'Mémoire sur la monarchie des Mèdes', Gibbon reflected on how authors stamp their authority upon their own creations:

> Every man of genius who writes history infuses into it, perhaps unconsciously, the character of his own spirit. His characters, despite their extensive variety of passion and situation, seem to have only one manner of thinking and feeling, and that is the manner of the author.

The comment obviously applies no less aptly to Gibbon's own *Memoirs*, a work of self-inspection so perceptive on the self that the notion of Gibbon as a 'one-book man' really must be abandoned: he was the begetter of a pair of masterpieces.

Gibbon's fascination with 'unconscious' processes (he uses that word himself, though more often his term is 'insensible') has of course been seized upon as a standing invitation by the psychohistorians, who have sat Gibbon on the couch and teased out his neuroses. Willis Buck, for example, tells us that Gibbon's attempts to turn himself into a Greek scholar as well as a Latinist was a sublimated bid to resolve his Oedipal conflicts. The 'full satisfaction' which he experienced with Rome represents his 'object-cathexis' with his (absent or dead) mother; but this was continually threatened, as Freud would predict, by his simultaneous need to identify with his father, symbolically expressed in his desire to 'master' the more authoritative Greek (Homer is 'the father of poetry'):

The resolution to the Oedipal conflict for Freud results from an intensified indentification [*sic*] with the father and the giving up of the object-cathexis of the mother. Gibbon attempts such a redeployment of energy when he tries to learn Greek at Lausanne.

When his father finally died, Buck explains, Gibbon's Oedipal desires could be permitted full play: he could identify with Mother Rome at last, and become its historian. Thus psycho-history makes short work of what had before seemed the rather intractable problems of explaining the historian's inner life.

All the same, Gibbon's consciousness also affords an intriguing story to us common-or-garden historians ('whatever else I may be ignorant of,' Gibbon believed, 'I think, I know myself'). He was intensely aware of how he had been driven into stories, romances and history by the lonely privations of his infancy. From childhood he had been captivated by *'un songe d'antiquité'*. But he knew as well that scholarship was not just a dream world, a present and even a future lived in the past, but a field on which he could exercise rational choice and judgment. The history of Rome came to him not as Oedipal resolution but only after laboriously sifting a range of subjects for their interest, importance and prospects for fame. Doing history was not just a matter of the lonely adolescent conjuring up a fantasy world which he could master and control, or a legitimate way for an adult to play with toy soldiers; it was the great theatre of mankind through which he could present the manifestations of virtue, power and wisdom, those leading concepts of the age, and their antitheses.

Gibbon introspectively examined that play of faculties and skills through which study and discipline crafted the often wayward – indeed perilous – enthusiasms of the imagination into historical truth; he tracked, through his own case, the forging of the historian's mind. But the public dimension of his vocation fascinated him no less than the inner. Was he worthy to tread in the footsteps of Herodotus and Tacitus? How would his work advance historical understanding? How might it secure him that 'immortality' to which it was the vanity of authors to aspire, the only kind of immortality for which an unbeliever such as himself could reasonably hope? Above all, how would it contribute to the mind of the age?

For Gibbon, history was worthless unless it constituted a school of virtue; it must be instructive. But it was not a mere pack of precepts, still less an oracle (prophets made his blood run cold). Establishing history's true place within the culture of the time, relating it judiciously to scholarship, arts and letters, was not the least of Gibbon's challenges and achievements. In the following chapters I shall explore how he came to write a kind of history special in both subject and form, and gauge his impact upon the outlooks of his age and upon the writing of history.

The best traditional Gibbon scholarship is straightforwardly biographical. The lives written some half a century ago by G. M. Young and David Low were not uncritical; both were, for instance, sharp with Gibbon over his rather shabby conduct towards Suzanne Curchod, the only person with whom he ever fell in love. But, writing at a time when the *Decline and Fall* still commanded a prominent place in every classically educated gentleman's library, these biographers laid their stress upon a life of achievement, and purred over its glories.

Now that the Classics have finally been devalued – even as I write Latin is being excluded from the core curriculum of British schools – the 'great work' may be becoming one of the great unreads. By consequence, though Gibbon scholarship multiplies, its focus has shifted. For one thing, the *Decline and Fall* tends to be approached differently. Not as a consummate example of the historian's craft, a model for the apprentice researcher to read and come away from filled with 'delight and despair'; but rather as a fascinating projection of the preoccupations, above all the tangled political ideologies, of the late Enlightenment. Recent students of the *Decline and Fall*, in particular J. G. A. Pocock, have been less concerned with evaluating its accuracy as Roman history than with teasing out its socio-political messages and meanings for Georgian England. The Gibbon who has recently come into prominence is the one who harmonized the contesting claims of freedom and authority, exposed the sociopathologies of fanaticism and superstition, and punctured the myth of the noble savage.

Relatively less, however, is now written about the *Decline and Fall* as such; far more about his *Memoirs*, in line with the belief, as Patricia Craddock phrases it, that we should 'portray the man as well as the historian'. This reorientation often arises out of the wish, born of Romanticism, to see the artist behind the art. In some cases it also stems from the quite different impulse of post-structuralist literary criticism to 'deconstruct' Gibbon the person and explore instead the textual representation of the 'author', inscribed in the seemingly endless proliferation of the *Memoirs* as an unreliable witness compelled to spin endless fictions about himself, an authorial voice woefully lacking in authority, relating a life dissolving into fragmentary discourse. Furthermore, in this age of psycho-history, we have a tacit expectation of finding answers to the Sphinx's riddle of creativity in the minutiae of the artist's everyday life – the sick infant whose wet-nurse lost her milk and whose mother died, the stubby, pudgy captain of the Hampshire Grenadiers trying so hard to be a man, the mute parliamentarian with, as Fanny Burney put it, 'those Brobdignatious cheeks', the historian so preoccupied with time that he became known as 'Mr Clockwork', the denizen of 'Fanny Lausanne' who posed as 'King of the Place'; the man who had a hydrocele – a gross swelling of the testicles – growing for twenty years without seeking medical advice.

Literary and psychobiographical studies addressed to these issues have been highly illuminating. Patricia Spacks in particular has stressed how the six jostling fragmentary lives stamp out Gibbon's dilemma; he was so anxious to write his life, yet – unlike his *Decline and Fall*, of which he sent the very first drafts of later volumes to the publisher – he found it almost impossible to accomplish. Similarly, Lionel Gossman has demonstrated how the problematical nature of legitimate authority in Gibbon's vision of imperial history was sharpened (to say the least) by his protracted attempts to come to terms with the imperious power wielded over him by his own father.

Too often, however, such psychobiography loses its sense of proportion. Thus Gossman goes on to assure us that Gibbon was 'crippled from the beginning by his family and his upbringing', hinting that he was either impotent or homosexual, and concluding that he 'resembled ... the eunuchs he pursued so unrelentingly in the pages of his great work'; for Gossman the *Decline and Fall* was 'the only achievement of a life of failure and renunciation'. But this is piffle. Gibbon had many notable achievements, not least several intimate, rewarding and enduring friendships, which demolish the diagnosis of him as an emotional cripple. He also, as we shall see, regarded himself not as a failure but as a thoroughly happy man. In his wisdom the doctrinaire Freudian triumphantly treats these 'denials' as confirmations: thus Gossman tells us that Gibbon's Augustan 'confidence ... displays ... the failures and anxieties it cannot quite conceal'. Above all, such psycho-history runs the risk of abandoning facts for fantasy. Gossman for example offers an interpretation of Gibbon's view of political power (we are told he hankered after undivided authority) wholly at odds with the historian's own often stated preference for a mixed constitution.

Another current of research has also moved away from the *Decline and Fall*, towards Gibbon's earlier ventures into scholarship. Some of these minor works were published during his lifetime, such as the fascinating yet neglected *Essai sur l'étude de la littérature* (1761), a work which almost launched the bilingual Gibbon into a career of publishing wholly in French; the two volumes of the *Mémoires littéraires de la Grande Bretagne* (1768–69), the review Gibbon wrote jointly with his Swiss friend, Georges Deyverdun; and the combative and anonymous *Critical Observations on the Sixth Book of the Aeneid* (1770), directed against the veteran Bible scholar, Bishop William Warburton. Others saw the light of day only after his death, in the collected edition of his *Miscellaneous Writings* issued by his faithful friend, admirer and literary editor, John Holroyd, Lord Sheffield. Amongst these are the chapters, in French, which Gibbon wrote towards his uncompleted history of liberty in Switzerland. Other writings, in particular his *Journals*, have only recently been published; still others, such as his various commonplace books, in which he docketed his working notes, remain unpublished in the British Library and elsewhere, testimonials to the value of a tidy mind and a well-organized data-retrieval system for the apprentice historian (Gibbon filed his notes on playing cards). The erudite analyses of this corpus of early writings offered by Patricia Craddock and David Jordan have demonstrated how the *Decline and Fall* did not spring, like Pallas Athene, full grown from Gibbon's bulbous red-haired head. Rather its form as well as its subject – indeed the very act of writing history itself – emerged only after a prolonged and often painful self-training. Gibbon entertained no Romantic vision of the author as spontaneous genius. When describing his art, he used the idiom of the artisan: 'the manufacture of my history required a various and growing stock of materials.'

Gibbon hugely admired Henry Fielding's *Tom Jones*; it was a work, he forecast, in one of his rare predictions, which would outlive the Escurial Palace. So perfect is that novel that it would be easy to assume that it came at the climax of a long tradition of novel-writing, stretching over many generations. Not so! We might easily be seduced into making the same assumption about Gibbon. So effortless, so assured, so natural appear his sweep, his design, his authority that we might take it for granted that Gibbon was simply the next in line to a long dynasty of English historical giants, indeed perhaps the last scion of a noble race, before the advent of the professional, professorial pygmies. But nothing could be further from the truth. Gibbon was a truly exceptional historian. How and why that is so will be explored in the next two chapters.

1 The Uses of History in Georgian England

'The eighteenth century is not usually regarded as an historically minded age,' wrote Christopher Dawson some fifty years ago, raising a canard about the 'age of reason'. It had been a charge levelled by nineteenth-century critics in particular, who condemned the thinkers of the Enlightenment for their aversion to a past from which they longed to escape, for their preference for the timeless over the time-specific, and for judging all ages by the supposedly eternal values of reason, nature and Antiquity. The mind of the *philosophe*, fine-honed for criticism, was accused of indifference to that truly historical understanding achieved by nineteenth-century historicism. It failed to grasp that each epoch had its own identity and that (as Ranke's dictum had it) all ages were 'equal in the eyes of God'; it lacked the urge to get under the skin and to enter into the heart and soul of the past. Enlightenment historians had sat in judgment (judged the Victorians), when they should have understood and sympathized.

Nowadays things seem less simple. We ourselves have questioned the shibboleths of nineteenth-century historians who thought they had discovered the secret of true historical-mindedness. Yet the Victorians' relativism itself became a new dogmatism. And having demythologized the Victorian myth of history, we now see the functions which such visions served, lending weight to the voice of the historians themselves, and promoting an image of Victorian society as rooted in the past while yet forging the future. We no longer take nineteenth-century historians at their word in claiming that the key to history was 'progress'; we see a mix of enduring structures and discontinuities where a century ago all seemed evolutionary. Hence there seems little point in judging eighteenth-century theory and practice of history-writing by the standards of the nineteenth. It is better to see how Georgian history served its own times and related to its own precursors, the investigations of scholars and antiquarians from the early Tudors onwards. We may conclude that the eighteenth century was not unhistorical, but that it had historical viewpoints of its own.

History-writing underwent great changes in England over the centuries from the Renaissance to Gibbon. By later standards, the tales of saints and chivalric heroes printed by Caxton and his successors still uncritically jumbled fact, fiction and fable; or, as Gibbon put it, 'to comply with the vicious taste' of his readers, Caxton 'amused popular credulity with romances of fabulous saints'. Seventeenth-century antiquarians, working in the tradition of Leland and Camden, often condemned their Tudor forebears as hopelessly credulous, or at least as lacking in the critical techniques needed to sift truth from propaganda; and eighteenth-century historians for their part belittled antiquarianism as myopic and pedantic. The antiquarian, it seemed to Gibbon amongst others, had learned to separate fact from fiction only at the cost of forgetting to distinguish between the important and the trivial.

Nevertheless it had been the crowning achievement of Renaissance humanism to generate a 'sense of the past'. The very act of playing off the luminaries of Antiquity against the barbarians of the 'Dark Ages' forced scholars to see difference, time, transformation and periodization. Thus the humanists' attempt to identify with the Ancients paradoxically drove them to formulate the novel concept of anachronism; and innovations such as printing, gunpowder and the compass, the discovery of the New World, and the New Learning of the Scientific Revolution, all stimulated new vistas of innovation and the race of time.

This is not to say that the cultivation of history presents a picture of steady improvement. Few eighteenth-century devotees could match the gargantuan learning of such Stuart scholars as Coke, Dugdale, Selden, Spelman, Brady, Hickes and Wanley, or even show a superior grasp of the development of the common law, the constitution or the feudal system. Perhaps no Enlightenment historian possessed so much information about the history and language of the Celts as the late-Stuart Welshman, Edward Lhwyd; and not until the nineteenth century did any scholar trace the fortunes of popular customs and beliefs with such curiosity and assiduity as had John Aubrey. The ferocity of political, constitutional and religious struggles in the Stuart age had spurred a quite extraordinary stampede 'back to the sources' in search of precedents and legitimations; and in many fields these labours were not surpassed till the emergence, in the Victorian age, of the disciplined cooperative endeavours of printing clubs, local history groups and archaeological and antiquarian societies – not to mention the rise of professional academic history itself.

Historians of historiography have sometimes claimed to espy 'revolutions' in historical thinking in the early modern centuries; be that as it may, it would be a mistake to interpret the course of historical studies in England from the Renaissance to Gibbon principally in terms of technical 'improvement'. It would be anachronistic to apply today's criteria governing good academic research to most cultivators of the past in earlier centuries. For one thing, most of them were amateurs, pursuing history in their spare time, for instance as passionate genealogists or historical topographers. For another, most were not writing history for its own sake, but pressing historical cases, evidence or documents into service for purposes which had little to do with the disinterested love of the past.

It is misleading to examine historical writing without asking about its audience. Before historians started writing mainly for fellow historians, their work was targeted at a whole host of readers. The Casaubonian pedant who obsessively amassed erudition, for its own sake or merely for the mice, was a perennial butt of humanist satire; and most historians took pains to ensure – witness the dedications to their books – that their work appealed to known consumers, be they local enthusiasts, political partisans and patrons, or the educated, leisured, *Spectator*-reading public. History-writing claimed, and perhaps earned, its keep precisely by presenting itself as useful, capable of teaching by example or of providing ammunition for the political and religious controversies ceaselessly waged in Stuart and Hanoverian England.

Unlike today's academic scholars, Georgian historians did not try to seal off their findings from the concerns of the present: quite the reverse. Certainly they laid claims to 'objectivity', but they did so not to insulate themselves from the ideological fray, but to enhance their credit within it. They all contended that the present must learn from the past. This was not new. Tudor humanists had commended their history as politically useful, offering lessons about the fortunes of rulers or presenting 'maxims' for the guidance of the commonwealth. In his Machiavellian *History of Henry VII* (1622) Francis Bacon relished offering such 'politique' advice.

Later, during the struggles culminating in the Civil War and the reigns of James II and William III, the past was pressed into service as offering solutions to constitutional conflict by revealing the *origin*, and thus the authentic *nature*, of institutions, titles, offices and privileges. Did the law take precedence over the King? Or did the King come before the law? Such questions could clearly be taken as prudential or metaphysical; but they could be answered historically as well. Vast quantities of scholarly time, ink and zeal in Stuart England were expended poring over charters and rolls and interpreting events such as the Norman Conquest, with a view to establishing the prerogatives of the sovereign under the Anglo-Saxons or the Normans, the origins of the common law, and the original powers of Parliament within the 'Ancient Constitution'. The governing criteria were precedence and priority. What came first was right, so the historian's business was to discover what came first. It was only the exceptional thinker, such as Thomas Hobbes, who queried the binding authority of antiquity in determining political rights and obligations.

Under the Stuarts and the Hanoverians, political argument assumed that the answers to, or at least the ammunition in, the great questions of the day lay in the past; English political battles were thus fought over rival readings of history. By Georgian times, practically all English history-writing was identifiable first by its dynastic and then by its party allegiances. This was one consideration, as we shall see, which dissuaded Gibbon from tackling modern history. History also provided the idiom for politics. Whenever Wolsey figured in an eighteenth-century cartoon, everyone knew the real target was not, of course, the cardinal but Sir Robert Walpole; Julius Caesar was an emblem for the Pretender in many a political pamphlet.

From the Reformation itself right through to the intricate ecclesiastical wrangles of the eighteenth century, questions of Church government were no less furiously contested via rival visions of the history of Church–state relations, thus stimulating monumental research, in Britain as on the Continent, into Christian history. As Owen Chadwick states,

> The Renaissance, the Reformation and then the Counter-Reformation gave that impetus to historical studies which begat modern historical writing. The origins of Christianity were in controversy. The nub of the argument between Catholic and Protestant lay in early Christian history.

History had been vital to Protestants right from the beginning, because the old Catholic question was so embarrassing: where was your Church before Martin Luther? Protestants could not duck explaining why Providence had permitted most of the history of Christianity – the period from the emergence of the Papacy through to the Reformation itself – to be the annals of ecclesiastical usurpation and doctrinal error, even, indeed, the reign of Antichrist. Protestants required a historical vindication to justify the otherwise rather mysterious ways of God to man.

The mainstream Protestant argument, one incorporated within Anglican apologetics, ran that the Roman Catholic Church was not a divine institution at all, but merely the bastard child of secular time and fallen man; thus the legitimacy of Protestant Churches depended upon recovering the essence of the original 'primitive' Church. But when, where and what was that? In Britain as on the Continent, the history of Protestant dissent proved a tale of claim and counter-claim as to precisely what the true, uncorrupted Church had been. Was it the gatherings of the Apostles, a congregational model? Or the Churches of the age of the Fathers, a more episcopalian pattern? Should the Jewish priesthood of Old Testament times serve as a blueprint for Christians? Or did the Erastian settlement established by the Emperor Constantine define true ecclesiastical polity in perpetuity, protecting the Church with the arm of the state? Catholics of course rejoined that Constantine had no sooner converted to Christianity and established it as the imperial religion than he 'donated' certain temporal powers to the Pope; but humanist scholars such as Lorenzo Valla and later Protestants had long since exposed this 'donation of Constantine' as a gross forgery.

In belief as in questions of Church government, orthodoxy seemed to hang upon disputes about the development of doctrine in the councils of the early Church. Throughout Stuart and Georgian England, battles over the history of theological formulae punctuated the polemics between latter-day predestinarians and free-willers, Arminians and Calvinists, Trinitarians and Unitarians. Thus prescriptivism dominated religious as much as secular politics. The search for an ideal, usable past which had agitated Henry VIII and Archbishop Cranmer reverberated through the Civil War, down into the squabbles between non-jurors, Jacobites, High Church prelates, Erastians, Latitudinarians and Dissenters under the Hanoverians.

Thus ecclesiastical conflict proved the spur, over the course of a couple of centuries, for a vast output of polemical history and minute grubbing into the archives of the past. But history also came to play a crucial role in a much more fundamental controversy: the question of whether the Biblical story of man, past, present and future, was the true account of Creation, or whether reason, science and nature had a quite different tale to tell about man, his nature and his past. All hung on the status of the Bible as a historical document. In England the issue was most sharply raised by Deists or free-thinkers such as Blount, Toland, Collins, Woolston and Tindal, around the beginning of the eighteenth century. Some of these saw themselves as purified Christians, some as defenders of natural religion, and some were probably sceptics: their subterfuges make their true beliefs difficult to pin down. These religious rationalists in effect battened on to the Protestant use of history and drove it one stage further, beyond where Anglicans were willing to follow; Deists used history to shake the historical claims of revealed religion.

If, as Protestants claimed, the history of Christianity – from some 'fall' in the early Church through at least to Luther – was largely the history of corruption, might not the *whole* of historical, positive Christianity be at best an irrelevance, at worst an error? If, as sober Protestants agreed, Judaism had been marred by ceremonialism, and Popery by superstition, how could we be sure that the doings of a few tiny communities of believers in the generations after Christ were intended to define true religion? Would it not be better to consult reason, conscience and nature, and to allow a generous tolerance in matters of creed and worship?

The cutting edge of the Deists' case was provided by the battery of historical arguments, long trained by Protestants against popish abuses, now redirected against those elements of historical Christianity which the Church of England defended. As Conyers Middleton implied in his *Free Inquiry* (1749), a work crucial to Gibbon's intellectual development, it was risky for Protestants to swing the blade of history against the papists, for it was double-edged. The thrust of the Deist argument was that positive religion was essentially a history of error, indeed of absurdity and imposture; far better to ground religion upon reason and conscience.

Thus alongside the orthodox use of history to legitimate the present by the weight of the past, a new scepticism arose, eating away at the historical foundations. In their different ways both the Renaissance and the Reformation had undermined the old certainties. The discovery of the New World and the rise of the New Science challenged the infallible authority of the Ancients, showed that not all truth was contained in canonical books, and dispelled the 'golden age' myth of pristine perfection. Descartes in turn had taught men to doubt, and Locke and fellow English empiricists argued that knowledge depended upon the witness of the senses and could never be more than probable. Overall, as Paul Hazard magnificently demonstrated in his interpretation of the early Enlightenment, the late seventeenth century bubbled with critical spirits such as Pierre Bayle, dedicated to sifting fact from fable in fields as various as ancient and exotic history, textual analysis, chronology, philology and mythology. The 'free inquiries' of these Pyrrhonists or sceptics challenged the authority of the records upon which the Christian world-view was built.

Deism struck at the roots by questioning, directly or by implication, the authority of the Bible and other sacred texts. It asked: wherein lay the authority of the Scriptures? Was it because they were 'inspired'? But in that case, by what historical, rational or scientific tokens was genuinely inspired Revelation to be differentiated from bogus, pretended revelation (e.g. the Apocrypha, or the holy books of other faiths)? If its credit rested upon its 'authenticity', i.e. the claim that it was truly of extreme antiquity, and that its accounts of events and peoples were capable of corroboration, then the Bible should, like every other document, be capable of independent corroboration – or falsification – by regular historical methods, by testing its congruence with the narratives, chronologies, myths and languages of other nations. Throughout the Enlightenment, the 'evidences of Christianity' were therefore subjected to intense scrutiny, by friend, foe and 'candid inquirer' alike.

To note these challenges from sceptics and Deists is not to suggest that religion itself was being shaken to the foundations from the late seventeenth century by a great blast of atheism, or that secularization was entrenching itself. All the same, blows were being struck at the pillars of a faith which had traditionally supposed that the Bible contained the truth, the whole truth and nothing but the truth (if properly interpreted). Protestant apologists had taken it as axiomatic that because the Bible was both inspired and ancient, its teachings were beyond question; it spelt out a comprehensive account of the world since the Creation, the providential history of man within it, and the rise and fall of the great empires, reaching forward, as prophets predicted, to the eschatology of the Second Coming and Last Judgment. Free inquiry in the humanities and sciences was not exactly making such views untenable, but it was demanding that defences be reinforced.

From the late seventeenth century onwards theologians tended to beat an orderly retreat from the claim that Scripture was a comprehensive scientific textbook of the natural history of the cosmos. Sacred history and natural history were slowly, sometimes agonizingly, sometimes easily, decoupled. The eighteenth century saw the triumph of theories in astronomy, physics and geology which were essentially naturalistic, and independent of Scripture. Genesis and geology went their own ways long before Darwin.

But the sacred history of man was a matter of infinitely deeper concern to Christian scholars. Starting around 4000 BC with Adam and the Paradise garden, and tracking, via Noah and Moses, up to Christ, this could not so lightly be relinquished, nor even treated as figurative; for the theology of the Cross, Resurrection and salvation hinged upon the truth of the Christian concept of time as an arrow. If history instead went in endless cycles, as the pagans claimed, it would make a mockery of the Crucifixion. Yet the authentication of the sacred history of man produced vexing problems. For one thing, distinct yet interlocking inquiries revealed the puzzles and paradoxes of the literal Scriptural story and chronology of the creation and dispersion of mankind, commonly believed to follow the Universal Deluge at the time of Noah. As scholars investigated Egyptian, Assyrian and Chaldean sources, and then lighted upon the records of the Chinese and the Hindus, they came upon a multitude of quite incompatible renderings of the history of mankind, often claiming to be far older than Jewish culture (and antiquity after all had a presumptive truth claim). Profound problems arose over reconciling this plurality of cultures, mythologies, religions, languages and histories – whose diversity was so threatening to the axiom that the Chosen People had been the fountainhead of all truth, morality and religion. In Restoration England, for example, the comparative chronologies compiled by Sir John Marsham –

his works formed part of the reading of Gibbon's book-hungry childhood – indicated that, despite the traditional orthodoxies, Egyptian civilization antedated the Hebrew. Late seventeenth-century sceptics and, after them, Voltaire loved to imply that Jewish customs, far from being aboriginal, were but the sweepings of the cultures of the Middle East.

Moreover, alongside this unsettling scholarship, many speculative philosophies were advanced in the late seventeenth century and through the *siècle des lumières,* offering visions of primitive man radically different from the Biblical. Not least, the fact, fiction or hypothesis of a primitive 'state of nature' was widely touted. From Vico through to Rousseau and the 'conjectural histories' of the Scottish Enlightenment, it was proposed that mankind had originated not as a fully-fledged civilized being, created in Paradise, on the sixth day, in God's image, but in a state of nature. The race had begun asocial, even savage, such theories suggested; man's subsequent development – linguistic, cultural, moral, cognitive and not least religious – had proceeded not thanks to the guidance of Providence or Revelation, but out of his own dynamic potential and his interaction with the environment. Such speculative anthropologies were rarely explicitly anti-Christian, though many seemed to have unChristian implications, such as a high antiquity for man. Even so, they presented human and natural explanations of those aspects of civilized existence – religion, morality, government – which Christianity had explained as purely divine in origin.

Faced with the acid of the Enlightenment, with what Peter Gay has called the 'rise of modern paganism', Christian scholars did not surrender. Nor did they, like their Victorian successors, beat a tactical retreat by arguing that the Bible was not, after all, a history book but rather a source of spiritual guidance. Instead they traded blow for blow. Defenders of the faith brought all possible resources to bear – the data of anthropology, philology, science and so forth – to construct a comprehensive history of mankind, its religions, tongues and empires, congruent with the account of human origins, dispersal and destiny expounded in the Old and New Testaments.

The case for the defence rested on two planks: antiquity and inspiration. On the one hand, the attempt was made to demonstrate that the Bible was indeed the most coherent, comprehensive and ancient account of early human history. If the Bible could be proved beyond doubt to be an accurate document, then its claims to be

God's Word would be bolstered. And if this were sustained, then the Old Testament's narrative of man's providential destiny must in turn command acceptance as the witness of a trustworthy, indeed inspired penman, Moses.

For Christian scholars prophecy and miracle constituted the vital props. Prophecy formed a kind of experimental test of the inspiration of the Old Testament. If 'candid' inquiry showed that the events foretold had indeed come to pass as predicted, was not that positive proof of the prophets' divine inspiration? Similarly with miracles. If the Biblical miracles had been seen and recorded by reputable eyewitnesses of unimpeachable veracity, then, once again, both the divine nature of Christianity and the credit of the Bible as a historical source were at a single stroke confirmed. Vast quantities of 'evidential theology' within this genre were devoted in late seventeenth- and eighteenth-century Europe to the 'candid' examination of prophecies and miracles contained in Scripture, the writings of the Church Fathers and the lives of the saints, cross-examined by regular canons of evidence derived from legal and historical practice.

This scheme of history as the fulfilment of a divine plan remained powerful through Georgian England. It was not an outdated genre. On the contrary, leading intellects stinted no effort to bolster it. No less a scientist than Isaac Newton devoted years to using his knowledge of positional astronomy to prove the accuracy – suitably adjusted by Newton himself – of Old Testament chronology, thereby exposing the 'falsehoods' of the annals of the heathen Egyptians and Greeks. For Newton, as for many of his contemporaries, such conclusions had profound implications, for they established the priority of Hebrew over Hellene, sacred over profane, culture. One of Gibbon's youthful exercises was to amend the Newtonian chronology. Far from an aberration, it was, indeed, typical that the gigantic twenty-volume *Universal History* (1736–65) – another work which young Gibbon devoured – devoted most of its first volume to the history of the Creation, and its second to the early events of the Old Testament; perhaps as a consequence it did not stretch nearer to the present than the Renaissance.

Nor were such scholarly calculations the work of cranks or monks; often it was the most rational and liberal Protestants who led the endeavour. Possibly because they prided themselves on being free of a Church establishment which dictated authoritative doctrine, Protestants looked to history and science to buttress their faith. Protestantism moreover, battling against the popish Antichrist, had unleashed great expectations of the millennium – still a tangible prospect to Gibbon's antagonist Joseph Priestley. Thus the drama of God revealing Himself in time past and time future was central to Protestant thinking. It had been set out in epic form by Milton in *Paradise Lost*, and in sober scholarly detail in works such as Bishop Stillingfleet's *Origines Sacrae* (1662) and Judge Matthew Hale's *Primitive Origination of Mankind* (1677), and in Catholic France in Bishop Bossuet's *Discourse on Universal History* (1681) – yet another work which profoundly changed the course of Gibbon's life. Belief that God directed the providential flow of time from Creation to Judgment Day formed the eschatological framework within which universal history was cast. In turn it supported a view of man as an actor in the divine drama of the Fall and Resurrection. The proper study of mankind was Providence. As the much-read French Jansenist Charles Rollin argued in his *Ancient History* (1738), the annals of mankind – which he judged began with the 'dispersion of the posterity of Noah' – showed the realization of a single divine plan. Every page of history displayed:

> the precious footsteps and shining proofs of this great truth, namely that God disposes all events as supreme Lord and Sovereign; that He alone determines the fate of kings and the duration of empires; and that he transfers the government of kingdoms from one nation to another, because of the unrighteous dealings and wickedness committed therein.

Within this epic vision, individual works of scholarship attempted to resolve particular difficulties. Influential in England was Bishop Thomas Sherlock's *Tryal of the Witnesses* (1729), which sought to establish that, by the regular canons of historical evidence, the scripture writers were trustworthy eyewitnesses, and hence their reports of miracles demanded credit. It was this genre which provoked David Hume's notoriously sceptical essay on miracles, which argued that, faced with an alleged miracle, common sense must judge that a breach in the laws of nature is less credible than that a witness should lie or be deceived. Hume's essay made a powerful impact upon Gibbon.

No less influential was William Warburton, later Bishop of Gloucester, whose *Divine Legation of Moses* (1737) offered the most pugnaciously learned defence of Scripture history. Warburton contended that the fact that the Old Testament Jews did not advance belief in personal immortality, a heavenly reward, confirmed that the Judaeo-Christian religion was not (unlike other faiths) an imposture. Had the Jews been only pretending to be God's Chosen People, opportunism would have led them to concoct a theory of eternal rewards. Gibbon found Warburton's casuistry devious and was provoked to rebut him in 1770.

These visions of history as a sacred drama were the books which young Gibbon first digested. Within them, the history of Rome had its own meaning as part of Providence's comprehensive plan; it was, particularly for Protestants, the last of the four terrible, godless empires (following Assyria, Persia and Alexander's Macedonia) before the coming of the 'fifth monarchy', i.e. Christ's. Gibbon eventually became disenchanted with all this cooking of the chronological books and prying into Providence. 'The last century,' he remarked, 'abounded with antiquarians of profound learning and easy faith, who by the dim light of legends and traditions and etymologies, conducted the great-grandchildren of Noah from the tower of Babel to the extremities of the globe.' But he dismissed them precisely because he had first read them all, had once believed them, had battled with them throughout his sickly youth, and finally emerged (as he would see it) into broad daylight, with a new vision of a purely human universal history, which encompassed – for the first time – both politics and religion.

History thus occupied a privileged place in the post-Renaissance mind. This was partly because it was valued as surrogate experience, and thus as, in Bolingbroke's phrase, 'philosophy teaching by examples'; but, more importantly, the past seemed to hold the key to the present, through having laid down positive (albeit fiercely contested) and binding title-deeds of legitimacy: political, legal and ecclesiastical. Over time the uses of the past became modified, but its grip upon the present was hardly diminished. That was why, at the close of the eighteenth century, Tom Paine still felt obliged, in his *Rights of Man*, to free mankind from bondage to the past, to inherited power, property and privilege and quite specifically to the 'Norman Yoke'; and why, equally, Burke's conservative politics of prescription gained such a purchase, not least over Gibbon in his last years.

Yet these uses of history did not predominate to the exclusion of all others. In particular, the shape of political controversy changed in the Georgian age, requiring a new history. Once the Hanoverians finally established themselves in the saddle, the Jacobite debate over the title and legitimacy of the Georges eventually subsided. Gibbon's grandfather seems to have been a hearty Jacobite; his son in turn hankered after the Stuarts, only to abandon that lost cause for regular Tory politics; and Gibbon himself turned Whig.

Instead, what increasingly engrossed the political nation were the practical questions of how power should properly be exercised – and, above all, curbed – within the mixed constitution established after the Glorious Revolution of 1688. What appeared to many to be the pernicious extension of centralized executive power, first under Walpole and the Whigs, and then under George III and Lord Bute, called forth a succession of opposition *exposés* of the threats posed by 'despotism' and tyranny' to liberty and the British constitution. Oppositionists developed what has been called a political sociology, deriving from the mid-seventeenth-century thinker James Harrington, who in turn had drawn heavily upon the republican writings of Machiavelli. In this anatomy of power the despotic tendencies of central government were seen as precariously held in check by the independent property-holders of the political nation. The health of the body politic would be secure so long as the propertied classes remained politically independent and vigilant. Freedom was jeopardized if they lost their virtue, either through being seduced by the apolitical pursuit of riches and luxury, or by allowing themselves to be suborned by the lures of office and rewards (in the jargon of the day, 'corruption').

Those viewing politics through Harringtonian spectacles – and they included controversialists such as John Toland and Whig political writers such as John Trenchard and Thomas Gordon – aimed to clinch their analyses with cases drawn from history. They used the past not to establish binding legal and constitutional precedents, but to lay bare a certain political anatomy. Within this genre, as Pocock, Turner and others have demonstrated, the example of Rome and its transformation – or rather degeneration – from republic into empire, became a classic object lesson in political suicide. As the respected journal, the *Monthly Review*, put it in 1764:

It is certain, that a thorough acquaintance with the Roman

government must afford the most useful information to the subjects of a free state, and more especially to our own, for there is undoubtedly a very strong resemblance between the general forms of each; both being of a mixed nature, compounded of royalty, aristocracy, and democracy, though the respective powers of these three orders were, in each constitution, blended together in very different proportions. The fundamental principles in each, however, being so nearly similar, many profitable conclusions may be drawn from a comparison between the Roman state and our own; and from the fatal effects of party zeal, public corruption and popular licentiousness in the one, we may form probable conjectures with regard to the consequences which the same circumstances must produce in the other.

Notably in Walter Moyle's popular *An Essay Upon the Constitution of the Roman Government* (1726) – a book much used later by Gibbon – but also in Trenchard and Gordon's *Cato's Letters* (1720–24) and Edward Montagu's *Reflections on the Rise and Fall of the Ancient Republics Adapted to the Present State of Great Britain* (1759), Rome presented a fair prospect of a virtuous, flourishing, militarily successful commonwealth. At least, that was so long as she had remained a republic, animated by the political energies of a citizenry whose conflicts proved to be healthy safeguards against despotism, and free from the abuses of hereditary or arbitrary power.

By contrast, imperial Rome, its patricians corrupted by luxury, its plebeians slaves to bread and circuses, afforded a spectacle of arid despotism. The proponents of this vision of history feared that Britain, like Rome, would lapse from liberty into despotism through a chain of socio-political transformations set off by changing circumstances but exploited by crafty opportunists. The growth of the monied classes lacking landed roots would sap political independence. For paper wealth – typically derived from government office, patronage and perks, and in turn loaned to the government through the National Debt – would be utterly dependent upon central power. Commercial societies would be dragged into costly wars, leading to rocketing taxation and larger standing armies – developments which would further undermine the traditional political nation and concentrate power in the executive. Standing armies, dropsical bureaucracies, 'placemen' and hangers-on would render the nation helpless before the machinations of absolutism. Thus wealth would breed luxury, luxury corruption, and corruption despotism. Within this popular diagnosis – it has been called the politics of civic humanism – the past was viewed with nostalgia: the simpler society of the 'good old days' had presented fewer opportunities for overmighty central authority. The young Edward Gibbon probably heard his father and his cronies grumbling over their port along precisely these lines – and this despite the fact that Gibbon's family itself owed much of its upstart wealth to the extortionate profits made by Gibbon's grandfather as a government contractor during Marlborough's wars, and to South Sea Company speculations (though these monies were confiscated after the bursting of the 'Bubble').

Politically, this argument did not of course go unanswered. Most provoking, perhaps, was Bernard Mandeville's *Fable of the Bees* (1714), which argued that no matter how much people might rant against the evils of 'luxury', it was all so much hypocrisy. In reality, 'private vices' were 'public benefits'; so-called corruption really meant general enrichment, and strong government kept the peace. Whig apologists contended along similar lines that economic progress and political stability would between them secure peace, prosperity and 'politeness'. Such a view was given highly sophisticated historical form by the Scotsman David Hume. Analysing the sweep of events in Britain from the Middle Ages to the Hanoverians, Hume transcended narrowly legalistic and party-political allegiance. His concern was not whether the Stuarts or the Puritans had been right, but rather their respective contributions to the emergent British polity. Hume concentrated on the political realities of the times; for historical reasons, central power was indeed being consolidated, and this was not 'despotism' but rather political maturation, which would protect those fundamental freedoms of the security of property and capital. Constitutional checks and the tempering force of public opinion would safeguard individual liberty. Hume's *History of England* proved extremely influential. It raised the art of history above spiteful political partisanship, and showed that analysis of the interplay of men, events and impersonal forces could produce a more satisfying history than a polemic of heroes and villains. No wonder Gibbon put it down with mingled feelings of 'delight' at its excellence and 'despair' at being unable to match it.

It would be difficult to exaggerate the centrality of history to the

Georgian frame of mind. The leaders of the humanist culture the Georgians inherited and cherished (the fact that we call it 'Augustan' is itself highly significant) self-consciously acted out their lives on a historical stage, fortified by the maxims of the past, playing the parts of ancient soldiers and sages. It was utterly natural for Gibbon casually to refer to 'Achilles Pitt' and 'Hector Fox', or to dub his father a 'senator' during his parliamentary career. Antiquity, familiar through the Latin and Greek which formed the staple of every gentleman's education, taught morals and manners, and offered models for politics and patriotism, art and culture. In that sense, history was 'valuable', a chest of political wisdom, stuffed, according to Walter Moyle, with 'useful treasures'. The true value of history, argued Lord Bolingbroke, did not lie in learning for learning's sake – indeed, it would be folly to 'affect the slender merit of becoming great Scholars at the expense of groping all our lives in the dark mazes of Antiquity'. Rather it was didactic and utilitarian, providing aids to living, producing 'a constant improvement in public and private virtue'. Historical plays, such as Addison's *Cato* and Johnson's *Irene*, filled the theatres, and commercial art galleries were opened in the metropolis celebrating the heroes of history. The most highly prized genre of painting (one endorsed by Gibbon's close friend Sir Joshua Reynolds and by the Royal Academy) was history painting – the noble depiction of great events aiming to raise moral consciousness. Amongst the faculties which elevated mankind above the brutes, the humanist mind prized memory and its sister attributes such as fame, expressed in the obituary and the epitaph. What Dr Johnson dreaded more than all else was oblivion.

During the eighteenth century, hundreds of histories poured off the presses, many of them the work of eminent men and women of letters such as Tobias Smollett, Oliver Goldsmith and Catherine Macaulay. Yet – perhaps paradoxically, perhaps consequentially – the times were not propitious in England for profound and original historical scholarship. Hardly anyone – Gibbon of course excepted – either distinguished himself principally as a historian or has since won an enduring place in the history of historiography. The two other leading English-language historians of the age, Robertson and Hume, were both, significantly, Scotsmen – though fairness requires mention of pioneers in more specialist fields, such as Thomas Warton's account of the early history of English literature and Charles Burney's history of music.

How do we resolve this apparent paradox of a historically minded age producing little history that lasted? The explanation lies in part in unresolved conflicts about what history should be, and how profound learning and polite literature should interrelate. The turn of the eighteenth century witnessed formidable debates in England and France about the respective achievements of the past and the present, the rival merits of the arts and sciences, and the match between mental faculties and intellectual accomplishments. Though this 'battle of the books' or 'debate between the Ancients and the Moderns' eventually descended into cliché, it was of profound significance in orienting cultural priorities. In England, the advocates of Antiquity tended to be gentlemen of letters. Vocal amongst them was William Temple, a fierce partisan of the superior civility and wisdom of the Ancients. Unfortunately, Temple was not himself distinguished by learning; indeed, though a champion of Greek culture, Temple himself had more or less lost his own Greek (a rather embarrassing demonstration of his point about the degeneracy of the Moderns). In France, classical studies were firmly entrenched in prestigious learned societies: but the idolaters of classical Antiquity in England were chiefly private gentlemen: dilettanti, connoisseurs and art collectors, whose taste for ancient sculpture or for the newly unearthed treasures of Pompeii or Herculaneum often outran their erudition or judgment. The case for the Ancients was not supported by a crack corps of classicists.

For, by a peculiar irony, the best classicists in Augustan England sided with the Moderns in the 'battle of the books'. Notoriously, Temple's scholarship was torn apart by the superior learning of Richard Bentley and, after him, William Wotton. In a formidable display Bentley, friend of Isaac Newton and eventually Master of Trinity College, Cambridge, used modern scholarly techniques to prove that the epistles of Phalaris, believed by Temple to be the most ancient of the Greek writings, were a late Antique forgery. Bentley loved the Classics, but he owed his pre-eminence as prince of the scholars to his mastery of modern methods with texts. The same was true of his younger contemporary Wotton. This one-time child prodigy was an expert on ancient history, but he made great play of scientific critical techniques in his *History of Rome* (1701), above all advocating the study of medals, which he saw, rather like fossils, as exempt from the sceptical objections to which literary records were liable. He too was a Newtonian, proud that his own times were at the leading-edge in learning. Wotton and Bentley, both Whigs, who probably knew more than anyone about Antiquity, were temperamentally 'Moderns', precisely because it was the new scholarship which had opened up Antiquity to the understanding.

Allegiances like these gave rise to a curious paradox. Some of the most learned scholars in early Enlightenment England felt deeply committed to the ethos of the New Science – valuing facts, physical and artefactual evidence such as medals, scepticism towards received authority, and 'scientific' research methods, and being dedicated to the Baconian ideal of the advancement of empirical knowledge. These investigators were furiously energetic. Many wrote county natural histories which drew upon historical material; some, like William Nicholson, pursued archaeological studies; others, like Hans Sloane, eventually President of the Royal Society, collected fossils, manuscripts and antiquities. Many became fellows of the Royal Society and published on ruins and relics, on historical topography, on Roman remains and similar topics. William Stukeley, physician, clergyman, antiquarian and ardent Newtonian, surveyed Stonehenge and Avebury and researched into the Druids. Such investigators were prominent in the founding of the Society of Antiquaries in 1751, at significantly more or less the same time as Samuel Johnson published his *Dictionary* of the English language and the British Museum was established.

This antiquarian drive must not be belittled. Some of its erudition shone, though it had its bizarre and eccentric aspects. But what such information-gathering about the past rarely produced was *history*, if by that we mean the systematic analysis of evidence to recreate the life and times, the actors and actions, of the past. Nor did such scholars write for the polite world. Few people in early eighteenth-century England knew so much history as the Oxford antiquarian Thomas Hearne – as is evident from his admirable historical handbook, the *Ductor Historicus* (1704), which Gibbon read when young. But Hearne ('a clerk of Oxford', complained Gibbon, 'poor in fortune and indeed poor in understanding') signally failed to write great history, and ended up satirized by Pope:

> To future ages may thy dulness last
> As thou preserv'st the dulness of the past.

Why was all this so? It has been suggested by Joseph Levine that the intellectual protocols of the New Science bridled the historical imagination and sapped the narrative impulse. History had, after all, been downgraded in the psychology of creativity advanced by Francis Bacon. Bacon saw the intellect as a tripartite hierarchy. At its apex, reason was the engine for understanding causes, or in other words for the pursuit of science; beneath reason, there was memory, which was all that was required for the pursuit of history; while, lower still, poetry was feigned history, the work of unruly imagination. During the seventeenth century, the distinction between reason and the lower faculties became more heavily accentuated, and leading intellectuals such as Locke and Newton deprecated fiction and championed fact.

In such a climate, those emulating science and ardent for the Baconian advancement of learning felt primarily committed to precise empirical research. For such men, traditional narrative history was under a cloud, since it seemed to require little more than a knack of fine writing. Resembling thus a branch of rhetoric, it was perhaps the domain not of real scholars but of mere writers, even the Grub Street hacks. History was open to insult as an enterprise whose truth claims would hardly pass muster. For all his prizing of renown against the wastes of time, Dr Johnson, for example, had some pretty sharp things to say about the limits of historical knowledge ('Why, Sir, we know very little about the Romans'). Overall, his philosophy of history was simple and sceptical: 'That certain Kings reigned and certain battles were fought, we can depend upon as true; but all the colouring, all the philosophy of history is conjecture.' In other words, Johnson radically distrusted the historian's claim to resurrect the past. He could be a journeyman writer, one who daubed colour on, but he could be little more. Given these limited goals, historians did not even require much talent: 'imagination is not required in any great degree.' Excessive powers might produce dangerous results – thus Johnson condemned Robertson's work: 'it is not history, it is imagination.' In other words, viewed as a beast distinct from the scientific antiquarian, the historian might resemble a mere decorator or popularizer; yet the antiquarian enterprise did not lend itself to the construction of true histories, in the sense of reconstructions of the activities of people in society.

Thus there was a profusion of antiquarian fact-gathering, and a hunger for history, but little marrying of original researches, historical thinking and literary talent. This point can be further illustrated by looking not at philosophies of history but at the social position of historians. Obviously historians did not operate in Georgian England as a coherent professional body; being a historian was not yet a 'career' for which one had special university training before embarking upon 'research', doing a PhD and sinking finally into a chair. History was even less well organized and rewarded than natural science. The Royal Society operated as a science lobby, but there were no specialist institutions for historians as such – though the antiquarians succeeded in organizing themselves into a society in 1751 and the Society of Dilettanti was likewise formed to foster taste.

Scholarship had traditionally been the province of gentlemen and amateurs, and this situation had hardly changed before the professionalization of the humanities as a result of university reform in the mid-nineteenth century. History did not become a specialist subject at Oxbridge until well into the Victorian reign, at approximately the same time as the Public Record Office was founded and the state began to publish the primary sources of British history. A budding historian such as Gibbon could not expect to find himself supported by a public apparatus of learning, unlike his modern successor. Even so, he found that England was peculiarly deficient in aids to scholars. Unlike most of the cities he visited on the Continent, London, he noted, 'is still destitute of that useful institution, a public library'; and so 'the writer who has undertaken to treat any large historical subject, is reduced to the necessity of purchasing, for his private use, a numerous and valuable collection of the books which must form the basis of his work.' Moreover he was put out that on his election to the only formal honour he ever held – Professor of Ancient History at the recently founded Royal Academy – the main consequence of his appointment was that he had to fork out twenty-five guineas for the privilege, in a kind of sinecure through the looking-glass.

Gibbon did not crave headed notepaper or an institutional affiliation, for in his view it was men and books that counted ('what is a council, or an university, to the presses of Froben and the studies of Erasmus?'). But he did seek stimulus. And in Gibbon's eyes, London, despite the pleasures derived from membership of the Literary Club, where he conversed with Johnson, Reynolds and Burke, was an intellectual wasteland; too often it meant 'crowds without company, and dissipation without pleasure'. In Paris it was easy even for a stranger to gain access to lively intellectual circles. On his visit in 1763, he readily met the Abbés de la Bletherie, Bartélémy, Raynal and Arnaud, conversed with de la Condamine and Duclos, and dined with Diderot, d'Holbach, de Guignes (expert on the Huns), and so forth; London offered no equivalent delights:

> I had promised myself the pleasure of conversing with every man of litterary fame [he wrote of London in 1758]; but our most eminent authors were remote in Scotland, or scattered in the country, or buried in the universities, or busy in their callings, or unsocial in their tempers, or in a station too high or too low to meet the approaches of a solitary youth.

England was indeed ill-endowed with formal facilities for historical study. Professorships of history had been founded at Oxford and Cambridge; but the history chair at Cambridge soon sank into rank jobbery, and significantly the massive recent history of Georgian Oxford has no chapter on 'Historical Studies' though separate sections are devoted to such fields as 'Hebrew Studies', 'Oriental Studies' and 'Antiquarian Studies'. The omission is not accidental, for few of the professors of history wrote any. Between them, John Warneford and William Scott, sitting on the Camden chair of Ancient History from 1761 to 1785, managed to publish just a single volume of sermons; while Thomas Nowell, the Regius Professor of History from 1771 to 1801, let off only a few squibs on college politics. The point is not, of course, that no historical activity was going on: for example, Sir William Blackstone as Vinerian Professor of Law at Oxford elucidated the history of the common law (Gibbon analysed his writings). Rather, no organized encouragement was given to training historians, promoting research or teaching students. Gibbon disowned the university which had done nothing to stir his mind:

> To the University of Oxford I acknowledge no obligation; and she will as cheerfully renounce me for a son, as I am willing to disclaim her for a mother. I spent fourteen months at Magdalen College; they proved the fourteen months the most idle and unprofitable of my whole life.

He never ceased to congratulate himself on his lucky escape; he could easily have turned into one of those drones who had so signally failed to teach him before sinking into deserved oblivion:

35

If my childish revolt against the Religion of my country had not stripped me in time of my Academic gown, the five important years, so liberally improved in the studies and conversation of Lausanne, would have been steeped in port and prejudice among the monks of Oxford. Had the fatigue of idleness compelled me to read, the path of learning would not have been enlightened by a ray of philosophic freedom: I should have grown to manhood ignorant of the life and language of Europe, and my knowledge of the World would have been confined to an English Cloyster.

Gibbon was deeply sensitive to 'the old reproach that no British altars had been raised to the muse of history' and aware that 'the other nations of Europe had outstripped the English in the progress of History'. There was truth in the criticism.

To say this is not to be anachronistic. For historical scholarship had already been set on firmer footings on the Continent, from the seventeenth century onwards. Crucial to this was the early modern equivalent of today's academic community: the regular clergy. Gibbon had nothing but contempt for monasticism and its vows of obedience, poverty and chastity. But he was generous enough to recognize that cohorts of Catholic clerics had long headed the organized pursuit of historical scholarship. It was the Bollandist monks who had developed diplomatics – the science of interpreting the form of official documents, essential for exposing forgeries. And it had been the 'Benedictine workshop' of St Germain des Prés which had produced 'the Benedictine folios, the editions of the Fathers, and the collections of the Middle Ages'; if, by contrast, Gibbon continued, 'I enquire into the manufactures of the monks at Magdalen, if I extend the enquiry to the other Colleges of Oxford and Cambridge, a silent blush, or a scornful frown, will be the only reply'. Zeal, selflessness, dedication and cooperative endeavour – by encouraging these, the Catholic Church had supported a corps of nonpareil scholars at a time when, for example, the researches into Islam of Simon Ockley in Cambridge (he had published a pioneering *History of the Saracens*, which Gibbon respected) had led him first into debt and then into gaol.

The French state had also helped to support historical scholarship on a scale unknown in England. The Stuart monarchs had funded the post of Historiographer Royal; but it was often treated as a sinecure – Dryden was one of the recipients – and did not do much for scholarship. In France, by contrast, Louis XIV had set up the Académie des Inscriptions, a prestigious society dedicated to the preservation and study of historical remains, whose subsidized journal, one of the earliest learned publications in the humanities, gave Gibbon great pleasure:

> I cannot forget the joy with which I exchanged a bank-note of twenty pounds for the twenty volumes of the Memoirs of the Academy of Inscriptions; nor would it have been easy by any other expenditure of the same sum to have procured so large and lasting a fund of rational amusement.

It is hard to imagine any publication produced by a learned society in Britain which would have kept Gibbon so happy. Though he kept up with science through the *Philosophical Transactions* of the Royal Society, there were hardly any specialist periodicals in the humanities.

The support given to historical scholarship by clerical learning and by the cultural aspirations of Continental absolutism had significant consequences. It encouraged the new standards of erudition and research techniques being developed in France, Italy and even, a little later, the German-speaking states. Erudition was not wanting in England, but it was all too frequently scattered and wasted. Much of the learning of men such as Lhwyd and Hearne never reached print (contrast the cascade of volumes published by Muratori or Tillemont). Poverty, insecurity or isolation turned many English scholars into blighted and sometimes misanthropic eccentrics.

It is noteworthy that in the twilight of his career Gibbon was keen to promote a new project for printing scholarly editions of the texts of early English history, from Gildas to the Tudors. Gibbon – not nearly so hostile to medieval history or to its chroniclers as is often made out – saw that, without such editions, proper history was impossible. What is striking is that it had not hitherto been accomplished. Significantly, the project, launched by the Scottish antiquarian John Pinkerton, was to be financed not by Parliament, nor by the Crown, nor even by the universities, but by a consortium of booksellers. In England, the production of history was largely a commercial operation dependent upon the taste of the public and the enterprise of publishers.

Studies of the profession of letters and the book trade have shown how the production and consumption of culture were more market-dominated in England than elsewhere. Personal patronage declined in importance, as Samuel Johnson noted, and so to win success and fame aspiring authors had to tailor their works to the wants of the customer, or to stimulate those wants themselves. Literary historians have explained how public taste, diffused through new media such as literary magazines and circulating libraries, made its mark upon the genres and content of writing, as for instance in the rise of the novel. The same applies to scholarship too. Philosophers, biographers and historians writing mainly for a broad, literate audience had to adjust their sights to suit their readers. This was a matter of convenience and necessity; but it also became a matter of choice. For it was now argued that authors had all too often been dry-as-dust pedants, composing in a 'barbarous' style, with little sympathy for their long-suffering readers. Nowadays it was necessary, by contrast, as Addison and Steele urged in the *Spectator*, to bring learning out of the academy and the cloister and into the drawing room; history had to entertain, instruct, enthral. Gibbon endorsed these new priorities. We often find him saying 'the public is the best critic' or 'the public is seldom wrong'. One noteworthy feature of the *Decline and Fall* is his masterly capacity to converse with his readers.

This drive to produce works of scholarship which the booksellers would commission and the public would buy affected history-writing. Gibbon apart, there are two all-round success stories. William Robertson produced a string of works – the *History of Scotland* (1759), the *History of Charles v* (1769) and the *History of America* (1777) – which were clearly written for a wide audience. They covered vast and important themes; their style, though never scintillating, was direct and strong; Robertson wrote an able narrative, uncluttered by digressions or minutiae. His books proved best-sellers and he was amply rewarded for his labours. (*Charles v* made £4500, and earned Robertson a gold snuff-box from Catherine the Great.)

David Hume presents an even better example. Possibly the most subtle thinker in eighteenth-century Britain, Hume had produced early in his career his philosophical *Treatise on Human Nature* (1739). It had fallen 'dead-born from the press'. Thereafter, and possibly as a consequence, Hume dedicated himself to essays and above all to history. His *History of Great Britain* appeared in three parts between 1754 and 1761 (he worked his way backwards from the late Stuarts to the Middle Ages), sold well, bore good financial fruit and won Hume fame, not least for the polish of his performance. It was no accident that Gibbon admired Hume and Robertson above all other British historians (though, ever the scholar, he found some of Hume's work somewhat 'superficial'), or that he was tickled pink that his own work found a place 'on every table, and almost on every toilette'.

Robertson and Hume were, of course, the exceptions, and we remain lamentably ignorant about the mass-production of history for the common reader. It is clear, however, that history commanded a large readership, and sales lists, booksellers' records and circulating library catalogues show that Georgian shelves groaned with works of history, alongside those two other non-fiction favourites, biography and travel-writing. But rising public demand often led to history in the Grub Street fashion: workmanlike multi-volume money-spinning hack jobs, aiming to fill gaps on gentlemen's library shelves, designed to satisfy the reader's – or at least the buyer's – desire to get good 'coverage'. When Lawrence Echard's *Roman History* – another of Gibbon's youthful favourites – sold well, an anonymous author was commissioned to write a continuation. Another similar enterprise was the *Complete History of England* (1706; often reprinted), a work in three volumes, under the general direction of the minor professional writer, John Hughes (one of Pope's 'dunces'). At mid-century Tobias Smollett was responsible for producing another multi-volume *Compleat History of England* (1757–58); shortly afterwards, Oliver Goldsmith brought out his *Roman History* (1769) and his *Grecian History* (1774), aimed at the juvenile end of the market. These and many similar works were got up by penmen writing to commission; they involved digesting existing scholarship and imparting to it an agreeable varnish. They required no special expertise: Goldsmith, grinding away at his Greek history, is supposed to have asked Gibbon, 'Tell me, what was the name of the Indian king who fought Alexander?' 'Montezuma,' replied Gibbon, and Goldsmith innocently wrote it down.

Such projects commonly contained no fresh research, and were devoid of scholarly apparatus. They satisfied a public hunger for kings, heroes and battles, and popularized rather than advanced the cause of history. In pandering to public taste, they confirmed the equation of the historian with general commercial writers. It is no accident that when those immortal, if apocryphal, words were addressed to Gibbon, 'So, I suppose you are at the old trade again – scribble, scribble, scribble,' the historian was automatically identified with a Grub Street scribbler, and history was taken as a branch of trade. Gibbon would not have been surprised, for he thought that no English historian in his own century deserved elevation to the pantheon.

This chapter has addressed a paradox. Georgian England was a land saturated with history. But, despite the popularity of antiquarianism, circumstances were not conducive to profound works of historical research. History was largely put to use as the raw material in a vast variety of productions, from party pamphlets to ecclesiastical polemics, Haymarket plays and political cartoons and satires. The past was almost too alive, too 'present', to permit great histories.

This sets the scene for Gibbon himself. It helps to explain why this voracious young reader binged on works of history, both sacred and profane – as he recalled, 'many crude lumps of Speed, Rapin, Meseray, Davila, Machiavel, Father Paul, Bower etc passed through me like so many novels'. In his hard reading, Gibbon may not have been such an unusual child. But it does underline his real distinctiveness. He was most unusual in eighteenth-century England in dedicating all his energies to historical scholarship, and in producing a work of great originality. This he could achieve partly because he possessed the independent gentlemanly income to support leisurely research – indeed to build up his own private library of 7000 choice volumes ('my seraglio was ample'). As Gibbon was fond of reflecting, he was blessed with the happy Horatian mean of wealth enough 'to support the rank of a gentleman, and to supply the desires of a philosopher,' without such riches as would render him idle. No real threat of poverty deflected him from the calm sense of mission spread over the twenty years it took him to complete his 'great work':

> Few works of merit and importance have been executed in a garret or a palace. A gentleman, possessed of leisure and independence, of books and talents, may be encouraged to write by the distant prospect of honour and reward; but wretched is the author, and wretched will be the work, where daily diligence is stimulated by daily hunger.

But he could also achieve it only because he was a singular human being. He worked on his own. For long stretches 'my only ressources were myself, my books, and family conversations' ('but these were great ressources', he adds). All the same, in one crucial respect he produced, and prided himself on producing, history as demanded by, history as lauded by, his compatriots. For Gibbon wrote for the public. He wanted to be a scholar, but he wanted to be read – as Hume told him, 'your work is calculated to be popular.' Indeed it was, successfully marrying those qualities of profound research and literary flair which had so typically been distinct in Georgian England; it made Gibbon some £9000. The copies of his books on the table and the toilette counted for more than those abused by the Oxford clergymen who fired off their ripostes. Gibbon was an exquisite mixture of the singular man who was also *par excellence* a man of his times.

2 The Making of the Historian

In his *Memoirs*, Gibbon claimed that 'I *know* by experience that from my early youth, I aspired to the character of an historian.' This may be mere hindsight, another variant on the old myth of the hero, revealing his destiny from childhood. Certainly in other moods the historian's fate seemed far more fortuitous: 'I look back with amazement on the road which I have travelled, but which I should never have entered had I been previously apprized of its length.'

All the same there may be a certain truth in the recollection. If Gibbon's *Memoirs* give anything like an accurate account of the early years following his birth in 1737 – and, in so far as they can be checked against his letters and journals, they do seem more like fact than fantasy – he was captivated from an early age by the past, by exotic tales and romances. Alone, neglected ('I seldom enjoyed the smiles of maternal tenderness') and frequently sick, the little boy found adventure and solace in books. He relished the *Arabian Nights*, and developed that utter absorption in Eastern mysteries – the mid-century equivalent of Gothic novels – which never palled: 'the dynasties of Assyria and Egypt were my top and cricket-ball.'

Especially after the death of his mother in 1747 when he was nine, he was frequently left in the charge of his favourite maiden aunt, Catherine Porten, with whom he 'turned over many English pages of Poetry and romance, of history and travels'. Her home across the way in Putney, vacated by her father who had absconded as a bankrupt, boasted a respectable gentleman's library, through which the child 'rioted without control'; and, between them, young Edward and Aunt Kitty read their way through classics such as the *Odyssey* (in Pope's verse translation), as well as tomes of oriental history and Biblical scholarship:

To her instructions I owe the first rudiments of knowledge, the first exercise of reason, and a taste for books which is still the pleasure and glory of my life, and though she taught me neither language nor science, she was certainly the most useful praeceptor, I ever had.

In particular, Aunt Kitty taught him to love books: 'to her kind lessons I ascribe my early and invincible love of reading, which I would not exchange for the treasures of India.'

His own Putney home was full of religious scholarship, for it had once been the residence of his father's sometime tutor, William Law, the eminent non-juror High Church pietist and eccentric mystic. Samuel Johnson had had Law's works dinned into his head by his mother, and his terrifying vision of the Lord haunted Johnson all his life. The impact of Law's piety upon Gibbon was more oblique, and revulsion from his view of a punitive God later spurred Gibbon's religious scepticism. But growing up in a home reeking of sanctity instilled in the boy an enduring fascination with religion and that inquisitiveness about theology which later pitched him into so much trouble.

But matching his taste for religion, Gibbon also developed an early love for the Classics, albeit necessarily in English translation; they schooled his mind and heart:

> The verses of Pope accustomed my ear to the sound of poetic harmony: in the death of Hector and the shipwreck of Ulysses I tasted the new emotions of terror and pity, and seriously disputed with my aunt the vices and virtues of the Heroes of the Trojan War.

Amidst these 'Putney Debates' Gibbon plunged into classical and humanist history, reading Herodotus, Tacitus, Machiavelli and Sarpi in translation, as well as travellers' tales of China, Mexico and Peru. His early formal education was, however, a shambles. He spent a year at grammar school in Kingston – illness put a stop to that; at eleven he was enrolled at Westminster School, but once again sickness prevented regular attendance and meant that he lost the normal schoolboy drilling in the Latin tongue and literature (he never did acquire the 'ear of the well-flogged critic'). These interruptions, as Gibbon later recalled, proved blessings in disguise, for he basked in the 'sick role', using his time off school to gobble up all the books he could lay his hands on:

instead of repining at my long and frequent confinement to the chamber or the couch, I secretly rejoyced in those infirmities which delivered me from the exercises of the school and the society of my equals. As often as I was tolerably exempt from danger and pain, reading, free desultory reading, was the employment and comfort of my solitary hours.

It didn't teach him to think, he added, but it fired his imagination.

Having left Westminster, his early teens were spent partly at home, partly convalescing at Bath, and partly traipsing around after his father, now a desolate widower, trying by dissipation to blot out the loss of the wife he had adored. Young Gibbon lived *his* life in books: 'My litterary wants began to multiply; the circulating libraries of London and Bath were exhausted by my importunate demands, and my expences in books surpassed the measure of my scanty allowance.' His *Memoirs* record a fateful family visit to Henry Hoare at Stourhead. Hoare's celebrated landscaped grounds held no attractions for the bookworm; he secreted himself away in the library, burrowing into the *Continuation* to Echard's *History of Rome*: 'to me the reigns of the successors of Constantine were absolutely new; and I was immersed in the passage of the Goths over the Danube when the summons of the dinner-bell reluctantly dragged me from my intellectual feast.' It was a kind of epiphany. The meeting of barbarian and Roman of course dominated Gibbon's mature vision.

Gibbon was despatched to Magdalen College, Oxford, in 1751 at the early age of fourteen: what else was a despairing father to do with this strange precocious child who was all contradictions? 'I arrived at Oxford with a stock of erudition that might have puzzled a Doctor, and a degree of ignorance of which a school-boy would have been ashamed.' Oxford had a few pleasures. As a gentleman commoner, he was entrusted with the key to the 'numerous and learned' college library. He enjoyed striding over Headington Hill, chatting with his first tutor, Dr Waldegrave. He was fired by a desire for learning, an urge to marshal information, a passion for argument: these gave the solitary boy a sense that he was at least good at something. He lived through the lives of those about whom he read. In this the young Gibbon reminds one less of those other intellectual prodigies such as his Oxford near-contemporary Jeremy Bentham, or John Stuart Mill, than of the juvenile fantasy lives of budding novelists such as Jane Austen or Charlotte Brontë; Gibbon, it has been remarked, would have made an excellent novelist. The joy of books never palled; a letter from him as a middle-aged bachelor to his step-mother simply runs: 'Reading a quarter past nine Saturday evening. Tout est bien. E.G.'

But his learning still lacked method, plan and rigour. His Latin was patchy, his Greek exiguous and his mind directionless. Following his early fascination with the *Arabian Nights*, he investigated the Orient: 'before I was sixteen I had exhausted all that could be learned in English of the Arabs and Persians, the Tartars and Turks.' He had a whim to study Arabic. His tutor dissuaded him but, as Gibbon complained, did nothing to channel his youthful enthusiasm more profitably. After a year of haphazard reading, when he said he was 'enamoured of Sir John Marsham's Canon Chronicus', he 'resolved – to write a book!', one designed to rectify the history and chronology of the Egyptian ruler Sesostris. The work, it seems (Gibbon later burned his juvenile efforts), was a typical early Gibbonian fusion of proto-Romantic exoticism with chronological minutiae, in which he tried to prove that the King had been the contemporary of the Biblical Solomon. Gibbon was thus already deep in sacred history. The next year proved disastrously more wayward. He 'bewildered himself into the errors of the Church of Rome'.

It was a very easy snare for that immature, solitary reasoner to fall into, having long been immersed in those controversies over 'celestial politics' explored in the previous chapter. A great *succès de scandale* of the day was Conyers Middleton's *Free Inquiry* (1749). Middleton took up the debate about providential interventions in the early Church. Catholics had traditionally argued that the continuation of miracles into the early centuries of the Church demonstrated divine sanction for the Papacy, whereas Protestants generally discounted alleged miracles much after the age of the apostles. Middleton, who, though ordained into the Church of England, was a worldly, ambitious scholar, tinged with free-thinking, denied that there was any clear-cut difference in the credibility of early and late, 'Protestant' and 'Catholic' miracles. In effect – though Middleton did not spell this out in so many words – with miracles it was either all or none, and Middleton seemed to be plumping for none.

Pondering Middleton, Gibbon accepted the premise, but reversed the implied conclusion. Surely later miracles were, *ceteris paribus*, better attested than the earlier? If later miracles were credible,

Catholicism itself seemed confirmed. Gibbon thus converted himself on the strength of the 'evidences'. He travelled to London, was received into the Catholic Church, and informed his father of the deed in a letter oozing 'all the pomp, the dignity and self-satisfaction of a martyr'.

There was presumably in all this an element of theatrical adolescent revolt, to spite his father, his *alma mater* and everyone else who had neglected him. Yet it would be a mistake to belittle Gibbon's intellectual religiosity or his earnest quest for truth. Latter-day critics have denied Gibbon's religious temper, because they see in it little of 'spirituality'. But this judgment is anachronistic. As the previous chapter suggested, Christianity itself was seen to stand or fall not by feelings but by facts; for the young Gibbon, a fair trial of the facts had found in favour of Popery.

This act of folly proved Gibbon's salvation. Aghast and furious, his father exiled him, still only sixteen, in a 'state of banishment and disgrace' to Lausanne, to study under a Calvinist pastor, Daniel Pavillard, until he reconverted. It was clearly meant as a punishment, and was certainly experienced as such; Gibbon recalled his chagrin at his reduction from the status of a fashionable Oxford gentleman, with his suite of rooms and valet, to that of a mere schoolboy. He held out for a staggering eighteen months before he finally wrote to his aunt, at Christmas 1754, in an English tongue he was fast forgetting, 'I am good Protestant.' Gibbon was always intellectually tenacious and independent; he had read himself into Catholicism, and he would not recant till he had read himself out of it. During his Lausanne exile he read himself in fact into a hearty contempt for Catholicism and, probably, a scepticism towards Christianity itself. He lapped up authors such as Bayle who exposed theological sophistries, Fontenelle and others who mocked superstition, and Giannone, who excoriated 'sacerdotal power'. Nevertheless, his exile continued. His father kept him in Lausanne for a further three and a half years, with no visits home. Only when he came of age at twenty-one was he allowed back, and then specifically to sign legal deeds to rescue his father from the financial scrapes into which his fecklessness had landed him.

One of Gibbon's most endearing traits was his capacity to buckle down and make the best of things. He turned his exile into a 'fortunate shipwreck' ('such as I am, in genius or learning or manners, I owe my creation to Lausanne'). He had little freedom, and was allowed too little pocket money to go roistering with the young bucks passing through Lausanne on the Grand Tour (one escapade, when he lost a large sum gambling, taught him his lesson). So he got down to work. He learned French, of course, one of the many yawning gaps in his previous education; in little more than a year he was bilingual. Under Pavillard's patient direction, he made himself highly proficient in Latin, and mastered the taxing art of double translation, turning a passage first from Latin into French, and later rendering it back into Latin, testing its resemblance to the original. He also became competent, though never perfect, in Greek. The good pastor set him to translating the Greek Testament, but Gibbon convinced him that Homer would serve his education better, being free of 'the corrupt dialect of the Hellenist Jews'. At last, he records, 'I had the pleasure of beholding, though darkly and through a glass, the true image of Homer whom I had long since admired in an English dress.'

Above all, developing the 'salutary habit of early rising' and learning 'application and method', he now began systematically to master the writings of Antiquity, especially those of Rome. He revered Cicero in particular for the values his writings imparted: patriotism, courage, public spirit, manly freedom, friendship, self-knowledge and self-control. 'I read with application and pleasure *all* the Epistles, *all* the Orations, and the most important treatises of Rhetoric and Philosophy.' And then, Gibbon recalled in his *Memoirs*, after:

> finishing this great Author, a library of eloquence and reason, I formed a more extensive plan of reviewing the Latin Classics under the four divisions of 1 Historians, 2 Poets, 3 Orators and 4 Philosophers in a Chronological series from the days of Plautus and Salust to the decline of the language and Empire of Rome.

He adds, honestly enough, 'and this plan in the last twenty seven months of my residence at Lausanne [January 1756–April 1758] I *nearly* accomplished' (we can gauge that he read about 6000 pages of Latin during this period). Not least, from 1755 onwards, he kept a commonplace book of his studies 'according to the precept and model of Mr Locke'. It contains some 235 entries from at least sixty-two different sources, and is arranged alphabetically. Gibbon was already turning into a model of methodicalness.

Furthermore, he began the habit of interrogating and reviewing his materials. He tried composing critical essays ('reviews') on the authors he read, in a somewhat schoolmasterly way assessing their style, chiding their blunders, evaluating their intellects, and judging their place in the pantheon of authors. Above all he developed the art of studying with critical penetration, and fixing his sources in his mind. Here is part of his running commentary on Giannone's great *Civil History of Naples*, which he read in 1756, discussing one of the key intellectual props of Catholicism, the Donation of Constantine:

> M.Giannone solidly refutes the fabulous donation of Italy to Silvester, Bishop of Rome, which some writers pretend that Constantine made in 324, four days after having been baptized by him. It is demonstrated to be false by the following arguments:
> 1rst. Because neither Eusebius, nor the authors who have written the life of Constantine in so much detail spoke of it at all.
> II. Because Eusebius informs us that Constantine had himself baptized at Nicomedia a few days before his death (according to a bad custom rather frequent among the nobles of that century) and not at Rome in the year 324.
> III. Because we know by the dates of the edicts of Constantine that during the whole course of the year 324 in question, he did not set foot in Italy, but spent it entirely in Thessalonica.
> IV. But what proves beyond all doubt the falsity of this donation is that all the provinces of Italy remained subject to the successor of Constantine, who commanded them as their master until the destruction of the Western Empire.

Through exercising his faculties in little essays like this, Gibbon learned to cross-examine his sources, and discovered for himself that the historian need not be reduced to hopeless scepticism when faced with the imperfect records of time past. But matters of style and literary merit exercised him no less. During 1758 in particular, as he ploughed systematically through the Latin historians, Sallust, Nepos, Caesar, Livy, etc., he set himself the task of sizing up their strengths and distinct characters as authors. Thus discussing Livy, he proposes to:

> divide what I have to say into several sections; I shall have four: I. In the first, I shall say something about the person and work of Livy. II. In the second, I shall indicate some of the qualities that distinguish his history from most others. III. In the third,

I shall consider the objections and accusations that have been made against him; and IV. In the last, I shall make some detached remarks on some passages of this historian.

All his life, Gibbon thought knowing the historian was a vital prerequisite for knowing his work.

At the same time, Gibbon sharpened his scholarly skills on problems cast up by his reading. He developed a fondness for emending corrupt texts, that favourite pastime of classical scholars ever since the Renaissance. Indeed, from as early as his nineteenth year he was writing bold but capable letters to some of the more eminent scholars of Europe – such as Professor J. J. Breitlinger of Zurich – offering variant readings to iron out difficulties. Some were accepted; others were, rightly, turned down. Even from his student days he was eager to make a name for himself, and saw scholarship as his forte.

But Gibbon also pursued these interests for his own enjoyment. The better his grasp of the great texts of Antiquity, the more deeply he plunged into the murky pool of scholarship and commentary which had welled up since the Renaissance. Gibbon had an unquenchable thirst for exact information. He glutted himself on travel literature, perused geographies and natural histories, and pored over atlases and chronological tables (he always prided himself on his tenacious grasp of time and place). Before taking a trip with Pavillard around the neighbouring cantons, he boned up on Swiss topography, town history and the annals of the people.

His powers of concentration grew lethal. He liked querying the critics, probing their quarrels and exposing their ignorance. Questionable attributions, doubtful dates, blunders over chronology and geography – Gibbon delighted in catching his elders committing howlers, a pleasure he pursued later in his unpublished 'Index Expurgatorius', his catalogue of scholarly bungles. Unlike some historians, however, Gibbon outgrew this juvenile love of ticking off his betters, or at least turned it to humorous effect in the footnotes to the *Decline and Fall*, where he took benign pleasure in correcting the slips of those, such as Montesquieu and Voltaire, whom he truly respected.

As he gained scholarly stature, Gibbon also trained himself as a man of letters. He read the giants of the French theatre and made the personal acquaintance, though not exactly friendship, of

Voltaire himself, who at his château at Ferney, just outside Geneva, prided himself on staging private theatricals, mainly of his own plays. And, not least, he familiarized himself with the *philosophes*. The Enlightenment which Gibbon absorbed and enjoyed in his five years at Lausanne, and which ever after for him became definitive of 'good' philosophy, was that centring upon the free inquiries of the late seventeenth century and the first half of the eighteenth. It was the Enlightenment which came before the shift towards militant atheism, radical populist primitivism, materialism and moral and sexual relativism, those features we associate with the late Voltaire, with Rousseau, d'Holbach, La Mettrie and Diderot. The *philosophes* primarily available to Gibbon in mid-century tolerant Protestant Lausanne were rather the liberal critics who spoke most clearly to his own situation, particularly as one attempting to break free from, and later luxuriating in his escape from, the thickets of Catholic casuistry.

At Pavillard's behest, Gibbon had early immersed himself in Locke. He found Locke's empiricism, and his emphasis upon sense experience as the test of truth, invaluable in undermining his assent to Catholic theology. Was transubstantiation true? No; for if, as Locke asserted, truth rests upon the senses, these clearly found against the translation of wine and bread into the blood and body of Jesus Christ.

Alongside Locke, Gibbon read many of the great rationalist scholars of the seventeenth century, Pufendorff, Grotius and Pierre Bayle above all. Gibbon appreciated the parallels. The Frenchman Bayle had been brought up a Huguenot; like Gibbon, he had fallen a convert to Catholicism. Having reconverted to Protestantism, Bayle continued his religious odyssey, ending up as what he liked to describe as a true Protestant, that is one protesting against all creeds. Bayle's scepticism has been much debated: did he become a pious fideist, embracing a faith of learned ignorance, stressing the proud fatuity of reason's claim to plumb the authentic divine mysteries? Or was he a destructive sceptic through and through? Whatever Bayle's true convictions, his vast compilations, particularly in the *Dictionnaire philosophique*, of the contradictions of scholars and theologians breathed for Gibbon the pure spirit of intellectual freedom, and fortified a detestation of dogmatism which would last all his life.

Gibbon read Pascal in rather a similar light, and the *Lettres provinciales* became one of his perennial favourites. Gibbon hardly responded to Pascal's religious torment, his bewilderment beneath the hidden, unknown God. But he relished the Jansenist's ironical demolitions of those complacent Christian theologians, above all the Jesuits, who aspired to rationalize every nuance of the Divine Mind. Reading Pascal drilled Gibbon in the use of a 'grave and temperate irony, even on subjects of ecclesiastical solemnity,' as he later put it.

Close familiarity with Bayle – whom Gibbon thought 'had more of a certain multifarious reading, than a real erudition' – helped the young Gibbon get to grips with the astonishing diversity of customs, beliefs and habits amongst the nations. Bayle loved to contrast the preachings against the practices of the Christian Churches, and to juxtapose the myths and worship of other civilizations; the irregularities, extravagances and sports of the human mind were thus richly revealed. But instead of stopping short at presenting error, or even psychopathology, he edged towards a deeper anthropological understanding, a naturalistic vision of how customs and beliefs were best treated, not so much as true or false but rather as functions of different circumstances, societies and environments and their respective psychological needs.

Such an anthropology of belief was also prominent in other writers Gibbon read in Lausanne, not least Fontenelle and Fréret, both of whom explored the natural history of religion, or at least of pagan polytheism. Fontenelle focused on the psychology of oracles; Freret, like Hume, looked into the primitive roots of religion. In the state of nature, fear stimulated piety, so the argument ran; the need to cope with impotence in the face of a threatening environment begat polytheism, fetishism and superstition. For those who chose to read between the lines, the florid saint-cults of Catholicism could be seen as the real target of their allusive suggestions. The point was not lost on Gibbon, who portrayed Popery in the *Decline and Fall* as but old polytheism writ large.

Gibbon acquired a more secular education in relativism by reading Montesquieu. The latter's *Lettres persanes* had held up the mirror to the certainties of Christian civilization, deflating Eurocentrism and teaching readers not to take at face value the manners and morals of their own society; they must instead be seen as relative and functional (or dysfunctional). Not least, Montesquieu the stylist fired Gibbon's ambitions to excel in the French language.

If Gibbon's student notes show him posing in the guise of a scholar, he was indeed quickly forced to assess his commitment to the life of learning. In 1757, when he was twenty, he fell in love with Suzanne Curchod, daughter of a poor Protestant pastor from Crassy, just outside Lausanne: 'The love of study,' he told her, 'was my sole passion up to the time when you made me realise that the heart has its needs as well as the mind, and that they consist in mutual love.'

They wished to marry, and Gibbon agreed to seek his father's permission on his return to England. As he probably anticipated, young love was blighted. Gibbon senior, exasperated by what looked like yet another mad escapade by his son, proposing to marry a penniless foreigner, made the objections crystal clear. The tottering family fortunes certainly would not support young Gibbon as a gentleman scholar, with a wife and presumably a growing family, in the ease he had anticipated for the future. Marry and be damned, thundered the father: or might his son care to reconcile himself to a comfortable existence as a scholarly bachelor? Faced with this choice, Gibbon saw reason and capitulated: 'I sighed as a lover, I obeyed as a son.' We need not doubt that Gibbon sighed as a lover. But what he obeyed was not just filial duty but a prudent inner voice which told him that what he already loved above all else were his books and the command of his time. Gibbon never seriously contemplated matrimony again: he allowed no ties of the heart to jeopardize the freedom of his mind.

But what kind of vocation in letters should he pursue? As a result of his exile, Gibbon had turned into a strange creature, a displaced person: 'I had ceased to be an Englishman ... the faint and distant remembrance of England was almost obliterated; my native language was grown less familiar.' Though still in his early twenties, he had already exhausted most of the obvious career options. He might at one stage have contemplated a donnish life; but his Oxford débâcle had ruled that out. He had no passion for the law. A country parsonage, where he might enjoy the 'fat' – i.e. lucrative – 'slumbers of the church' while browsing through his books in peace, had also been eliminated. A man indifferent to country manners, company or sports, Gibbon could hardly look forward to further banishment as an antiquarian squire, rusticated in Hampshire. Lausanne had given him a life-long taste for the Continent. It might possibly have made him a home ('I should have chearfully accepted the offer of

a moderate independent fortune on the terms of perpetual exile');
but Lausanne was also small beer for an ambitious young man
with a name to make. London – where he hoped to 'divide the
day between Study and Society' – was, as Gibbon was soon to
find, costly, and high society was a cut above his class and pocket.
Paris was certainly inviting, with its special cocktail of fashion,
gentility and intellectual fizz; indeed, a few years later, in 1763,
'M. Guibon' seriously contemplated, for a moment, settling there.
But could he hope to be a literary lion in a foreign capital? And,
anyway, Paris too was expensive.

The questions of where to settle, and what to do to slake his
thirst for achievement, agitated Gibbon as he came of age. They
form the subtext to his first substantial piece of writing, the *Essai
sur l'étude de la littérature*, which he began in Lausanne, brought
back to England in 1758, revised, shut away in a drawer, showed
to friends, and finally published in 1761, thereby losing his 'litterary
maidenhead'. He did so, he later claimed, at his father's insistence
(perhaps we should take this with a pinch of salt), for a book to
his credit might help him, like Hume, to a diplomatic post, one
of the few attractive career options still open.

It is noteworthy, first of all, that the *Essai* was written in French.
It is not surprising that Gibbon should be thinking in French after
five formative years in the *pays de Vaud*, but it is revealing that
he targeted it to a Continental rather than a home readership. Indeed
he chose to write it in a scintillating, oracular style modelled on
Montesquieu, bursting with short, pithy sentences, each striving
for effect; the book was composed of brief chapters, arranged in
what looks almost like random order. Gibbon did not write the
Essai to establish his credentials as a dogged scholar toiling in the
vineyard of truth. He wanted to cut a dash and win his spurs as
a man of the Enlightenment, by playing variations upon some of
the *philosophes'* themes; this was a book to catch the eye.

And yet Gibbon chose not to salaam to the sultans of the Parisian
salons. Just the reverse. It has become fashionable, Gibbon
complains, to wave traditional scholarship aside as outmoded, dull
as ditchwater. Thanks to trend-setting publications such as
d'Alembert's 'Preliminary Discourse' to the *Encyclopédie*, being a
champion of erudition has become an embarrassment; such men
are sneeringly dismissed in the new jargon as *les érudits*. These
days what counts – Gibbon of course exaggerates wildly for effect –

is not deep learning but tossing around some glib philosophy.

All his life Gibbon found this *trahison des clercs* – this temper which led Voltaire once to dub 'details' as the 'vermin which destroy great works' – quite mystifying. In the *Essai*, he hints that Cartesian rationalism and the New Science should shoulder the blame. For these had introduced a taste for simplicity, clarity and novelty, and discredited Antiquity. The Cartesian Fontenelle has apparently dismissed ancient Greek history at a stroke as a collection of 'dreams and fantastic imaginings', 'a phantasmagoria, a string of childish tales'; and d'Alembert had floated the notorious suggestion that there should be a regular stock-taking of the annals of the past: every century the archives should be weeded out and the rubbish destroyed. Gibbon issued a sly riposte: there were no useless, meaningless facts – 'let us preserve them all most carefully. A Montesquieu, from the meanest of them, will draw conclusions unknown to ordinary men.' More broadly, the *Encyclopédie* created a climate lukewarm towards literature and history. D'Alembert had pinpointed three chief faculties exercised in mental labour: understanding and imagination, both of which were admirable, being responsible for poetry, oratory and science; and mere memory, which was the only faculty required for history and erudition. In d'Alembert's view, memory 'had been superseded by the nobler faculties of the imagination and the judgment'.

Against these polemics, Gibbon could hardly be expected to defend pettifogging pedantry, or mount yet another rearguard defence of Antiquity in the by then old-hat Ancients *versus* Moderns debate. He never had any sympathy for mere antiquarianism, so often trifling, partisan and crotchety; and certainly not for those who misused learning, such as Salmasius who 'too often involves himself in the maze of his disorderly erudition'. Instead Gibbon advances in the *Essai* a subtle defence of the true value of, and necessity for, critical scholarship, a more philosophical account of what scholars such as Bentley had been doing.

No one would deny, he argues, that *littérature* is the school of life. He took 'literature', a recently minted French word, to include not just *belles lettres*, poetry and rhetoric (though Gibbon champions these) but the whole corpus of the Classics. Classical learning presents man at his best: 'the ancient authors have left models for those who dare to follow in their footsteps.' Such works of intellect and imagination elevate mankind above the beasts and the cultivated above the mass. Literature civilizes; it teaches morals and manners; it instructs us in conduct and civic wisdom, stimulates the faculties, and gratifies our love of beauty. Of the world's literature, Gibbon proceeds, the finest body remains that in Greek and Latin. Even the most perfervid Moderns in the 'battle of the books' admit the superiority of the Ancients in the epic, in tragedy, moral philosophy and oratory.

All this was commonplace. Gibbon's novel point was to insist that literary works, in particular classical writings, wrenched out of context and read in ignorant isolation, lose their web of meaning. How can we truly appreciate the beauties of Homer without knowing about Greek history, religion and myth? Or understand Cicero without being familiar with Roman justice, or measure his political rhetoric ignorant of the Senate and Roman politics? How can we study Horace or Virgil without appreciating the conventions of Roman criticism? Abandon learning, and you sever the blossoms of taste, understanding and sympathy from their roots. Gibbon was scandalized to find the Academy of Inscriptions, that doughty guardian of erudition, 'degraded to the lowest rank amongst the three Royal Societies of Paris'.

The intelligentsia, Gibbon implies, will cut its own throat if it denigrates erudition and stakes all upon a kind of flashy wit. Proper criticism in fact helps reveal the true mission of the intelligentsia, past and present. Take the case of Virgil. For want of proper historical-critical insight, his *Georgics*, that celebration of rural repose, is commonly misunderstood. Perhaps with his tongue in his cheek, Gibbon fills in the background. Augustus needed to disband his victorious armies. Yet he was aware that, unable to pay the veterans their fat demobilization bonuses, he ran the risk of mutiny. By a stroke of artful genius – Gibbon always viewed Augustus as a crafty operator – he commissioned Virgil to write the *Georgics*. Their picture of the happy farmer, content in his rustic lot, won the soldiers over. They happily waited thirty years for their pay, and Augustus duly rewarded Virgil for his labours. All becomes clear:

> Virgil was not simply a writer describing rural labours. He was a new Orpheus sounding his lyre in order to calm the ferocity of savages and to unite them by the bonds of custom and law. His poem worked this miracle.

Thanks to learning, we can understand the crucial role that Virgil played – that letters played – in the Roman polity:

> It is impossible to appreciate the plan, the art, the details of Virgil unless we are thoroughly acquainted with the history, the laws, and the religion of the Romans, with the geography of Italy, with the character of Augustus, and with the peculiar and unique connexion which this Prince established with the senate and the people.

Using arguments, charming yet pointed, such as these, Gibbon tried, like a Renaissance humanist, to convince the learned world to value rather than despise its own forte: and the great world to imitate Augustus, and patronize letters and learning.

The *Essai* was a *tour de force*, even though it betrayed the inevitable immaturity of a first book, written by someone barely into his twenties. Gibbon was later dismayed at the artificiality of his style ('The obscurity of many passages is often affected, *brevis esse laboro, obscurus fio*'); yet it was a work upon which he could also look back with some pride. It did not, however, solve his most pressing problems. Put crudely, it won him some applause but neither a diplomatic post nor any other recognition.

It left him still dangling in the same unsatisfactory position in which he found himself on his return to England in 1758. His £300 a year allowance from his father, bought at the cost of jeopardizing the future through a broken entail, was enough to allow him to winter in London while grazing during the summers on the family estates in Hampshire. But neither environment was ideal. In London, where he took lodgings in Bond Street, he lacked the funds requisite to be a man of pleasure, and the introductions which would give him a secure *entrée* into the best society. 'While coaches were rattling through Bond Street, I have passed many a solitary evening in my lodging with my books': salutary mortification for the scholar, no doubt, but not much fun for a twenty-two-year-old just back from exile. In Hampshire things were even less satisfactory. The interruptions of family obligations piled on top of the calls of country life. Between 1759 and 1762, the demands on his time were dramatically increased by active service in the Hampshire militia – his father had signed up both himself and his son, never expecting the regiment actually to be commissioned; as ever, his judgment had proved wrong. Gibbon was on active service, marching yokels around the southern counties, between May 1760 and December 1762.

What was worse, young Gibbon, who lived at home throughout the 1760s with the exception of the *Wanderjahr* of 1763–64, found himself watching his father's extravagance, fecklessness and bad management eating up the family fortune day by day. Now moving into his thirties and chafing at the bit, Gibbon grew fearful that, if the father chanced to survive to a ripe old age, the son would inherit nothing but debts: perhaps even the gaol loomed before him, he reflected in one particularly melodramatic letter.

Gibbon toiled away at his studies. The journal kept in 1762 is a remarkable testament to his industry and zeal, and his capacity to keep chivvying himself out of the slough of despond and the Hampshire squirearchy; frequently drunk or hung over after family dinners and militia carousings, he dragooned himself back to Homer or Cicero. Keeping a diary would 'both ... assist my memory and ... accustom me to set a due value upon my time'. He loathed the interruptions. 'This morning was terribly broke into by the adjournment of our Court Martial which lasted 8 to 10, and a field day which I attended from eleven to one,' he records – though adding: 'However, as the greatest difficulties are those occasioned by our own laziness I found means to go thro' the Racines Grecques [his Greek syntax] from 28–32, to review the whole VIth book of the Iliad, and to read the VII v.1–123 in this busy day.' No side-tracking into such light reading as plays: 'I must learn to check these wanderings of my imagination.' Luckily scholarship made him happy, and he took pleasure in mastering the Greek and Roman calendars, even lying in bed at night reviewing their confusions. The serious, methodical little man kept plugging away:

> Designing to recover my Greek, which I had somewhat neglected, I set myself to read Homer, and finished the four first Books of the Iliad, with Pope's translation and notes; and at the same time, to understand the Geography of the Iliad, and particularly the Catalogue, I read the VIIIth, IXth, Xth, XIIth, XIIIth, and XIV books of Strabo, in Casaubon's Latin translation. I likewise read Hume's history of England to the reign of Henry VII, just published. *Ingenious but superficial*; and the Journal des Savans for August, September, and October 61, with the Bibliothèque

des Sciences, & c, from July to October 61 : both these Journals speak very handsomely of my book [i.e. the *Essai*].

Yet no settled direction emerged: Gibbon was still betwixt and between. His first love was literature, and so becoming a critical scholar was an attractive possibility. He also felt stirrings towards history, while doubting his abilities:

> My own inclinations [he wrote from military camp at Winchester] as well as the taste of the present age have made me decide in favour of history. Convinced of its merit, my reason cannot blush at the choice. But this is not all. Am I worthy of pursuing a walk of literature which Tacitus thought worthy of him, and of which Pliny doubted whether he was himself worthy? The part of an historian is as honourable as that of a mere chronicler or compiler of gazettes is contemptible. For which task I am fit, it is impossible to know until I have tried my strength; and to make the experiment, I ought soon to choose some subject of history which may do me credit, if well treated.

But what might have been the true epiphany merely led to a series of false starts. At the close of the Seven Years' War in 1763, Gibbon won his father's permission to spend money earmarked for putting him into Parliament on a Grand Tour instead. He basked in some acclaim in Paris, which perhaps made his obscurity back in England the more galling. The applause was marred only by his mild resentment at being taken primarily as a man of letters rather than a man of rank: Gibbon never lost his social aspirations. He met and dined with *philosophes*, and Madame Bontems condescended to flirt with him.

He then moved on to winter in Lausanne, installed himself in a hotel to escape Madame Pavillard's dirty tablecloths, and plunged into his books once again, not least fifty volumes of the *Bibliothèque Raisonnée*, a periodical published in Amsterdam between 1728 and 1753, containing reviews, abstracts and excerpts from scholarly publications. Above all, Gibbon prepared himself for what to an aspiring classical scholar was bound to prove the adventure of adventures: his first visit to Italy. He was as meticulous as ever. He systematically waded through all the standard sources on ancient Italy, in particular the historical topographies of Cluverius (*Italia Antiqua*, 1624) and Nardini (*Roma Vetus*, 1666). Not only did he read them, but, as always, he took copious notes, organizing his extracts by topographical location:

> From these materials I formed a table of roads and distances reduced to our English measure; filled a folio common-place book with my collections and remarks on the Geography of Italy, and inserted in my journal many long and learned notes on the *Insulae* and populousness of Rome, the Social War, the passage of the Alps by Hannibal &c.

In his journal he records that he conceived the idea of publishing a further book, a historical topography to be called the *Recueil géographique sur l'Italie*, organized regionally according to each area's role in the rise of Rome: 'The reader would then easily follow the progress of Roman arms and Livy's narrative.' The scholar in Gibbon rather fancied the idea: 'it could make a profit for a publisher' and be 'some use to the publick'. In the event, he never brought it out, though Lord Sheffield printed it posthumously. Written in French, it examines ancient Italy province by province, surveying the towns and their strategic and economic importance, and quoting poetic descriptions of them. It was a work precisely in line with the mission of scholarship – learning servicing literature – as set out in the *Essai*, and Gibbon's failure to publish suggests that the compilation of *Blue Guides* could no longer satisfy him: he was now setting his sights higher.

Gibbon could justly boast 'that few travellers more completely armed and instructed have ever followed the footsteps of Hannibal'. Carried over Mont Cenis on a litter, Gibbon had mixed feelings about the north. Turin confirmed for him all the evils of courtly absolutism, nor did Florence completely win him over – his sensibilities to the visual arts were never very vibrant. But Rome ('the great object of our pilgrimage') did. He arrived; he could not sleep, he wrote to his father, he was delirious as in a dream. As he recalled in his *Memoirs*:

> My temper is not very susceptible of enthusiasm, and the enthusiasm which I do not feel I have ever scorned to affect. But at the distance of twenty five years I can neither forget nor express the strong emotions which agitated my mind as I first approached and entered the *eternal City*. After a sleepless night I trod with a lofty step the ruins of the Forum: each memorable spot where

Romulus *stood*, or Tully spoke, or Caesar fell was at once present to my eye; and several days of intoxication were lost or enjoyed before I could descend to a cool and minute investigation.

Rome truly had its 'miracles'. Gibbon scholars disagree over whether:

it was at Rome on the fifteenth of October 1764, as I sat musing amidst the ruins of the Capitol, while the barefooted fryars were singing Vespers in the temple of Jupiter, that the idea of writing the decline and fall of the City first started to my mind.

But it is beyond doubt that the city, which he had imaginatively inhabited for so many years, affected him deeply. Even so, whether or not Gibbon first thought of writing the *Decline and Fall* at that time, the project did not drive all others from his mind. All that can be said with certainty is that from around 1764, the idea, not necessarily of writing the history of Rome, but of writing *history* as distinct from classical scholarship, antiquarianism, chronology or topography established itself.

But the history of what? It was a problem which perplexed Gibbon every time he contemplated a major *opus*. The biography of a great man had an initial appeal. Perhaps Charles VIII of France, or the 'crusade of Richard the first, the Barons Wars against John and Henry iii, the history of Edward the black Prince, the lives and comparison of Henry v' – all clear-cut evidence that it is wrong to think of Gibbon as indifferent to the Middle Ages. He also pondered 'the life of Sir Philip Sidney, or the Marquis of Montrose', or the attractive prospect of Sir Walter Raleigh. But even then he listed plenty of drawbacks in his journal, in a passage he subsequently thought important enough – as evidence of his bewilderment at the sheer plethora of possibilities – to include in his *Memoirs*. Raleigh's life had been adequately covered in Oldys's biography. So what did this leave him?

My best ressource would be in the circumjacent history of the times, and perhaps in some digressions artfully introduced, like the fortunes of the Peripatetic philosophy in the portrait of Lord Bacon. But the reigns of Elizabeth and James i are the period of English history which has been the most variously illustrated:

and what new lights could I reflect on a subject which has exercised the accurate industry of *Birch*, the lively and curious acuteness of *Walpole*, the critical spirit of *Hurd*, the vigorous sense of *Mallet* and *Robertson*, and the impartial philosophy of *Hume*?

Not least, how would he cope with the party-political bias required in modern history written for English readers?

Could I even surmount these obstacles, I should shrink with terror from the modern history of England, where every character is a problem and every reader a friend or an enemy: where a writer is supposed to hoist a flag of party, and is devoted to damnation by the adverse faction. Such would be my reaction at home: and abroad the historian of Raleigh must encounter an indifference far more bitter than censure or reproach. . . . I must embrace a safer and more extensive theme.

It is easy to appreciate Gibbon's quandary. A good biography of a home-bred hero such as Sidney or Raleigh, with its lessons for liberty, might have won friends and influenced people; it would have advanced his claims – strongly backed by his father – upon a political career. But the disadvantages were heavy. The inimitable Hume had already laid claim to the Tudors and Stuarts, Sidney and Raleigh were well-worn subjects, and Gibbon discovered there was little he could add. No use producing a work which would not crown him in glory.

That is why Continental subjects, to be penned in French, commended themselves more strongly. For scope, dramatic appeal and political ramifications, the 'history of the liberty of the Swiss' or, equally, the fortunes of Renaissance Florence both appealed. The Swiss subject had the great advantage of encompassing:

that independence which a brave people rescued from the house of Austria, defended against a Dauphin of France, and finally secured with the blood of Charles of Burgundy. From such a theme, so full of public spirit, of military glory, of examples of virtue, of lessons of government the dullest stranger would catch fire.

The history of Florence, by contrast, had the advantage of 'two delicious morsels', the revival of learning and the rise of the Medici.

Gibbon had been fired to start work on the history of the Swiss while in Lausanne in 1764. On his return to England, while he resumed his migratory life of wintering in London and summering in Hampshire, he pressed on, encouraged by the presence of his Swiss friend, Georges Deyverdun, who translated German sources he could not read. The dénouement is famous. Gibbon sent the first chapters of the manuscript to Hume, who was polite but not exactly aglow; Gibbon also arranged for it to be read anonymously in 1767 before a 'litterary society of foreigners in London':

> and as the author was unknown, I listened without observation, to the free strictures and unfavourable sentence of my judges. The momentary sensation was painful; but their condemnation was ratified by my cooler thoughts; I delivered my imperfect sheets to the flames.

He did not make a bonfire in fact, and the fragment was published by Lord Sheffield after his death. Gibbon's 'cooler thoughts' were right. It is a dull work, unrecognizable as coming from the same mind which a decade later would dazzle the world with the *Decline and Fall*. The Gibbon who wrote about the medieval Swiss was still far from being a fully-fledged historian. The writing is stylistically unsure; it lacks Gibbon's later flair for the set-piece, and his characterization was still rudimentary.

Moreover, Gibbon had shown poor judgment in his choice of theme in the first place. Unable to read German, he could never master even the printed sources and background materials, let alone explore the manuscripts. Obviously residence in Lausanne and comradeship with Deyverdun had swayed him. But an eye to the political attractions of the subject had doubtless also influenced his choice. Having come of age, Gibbon found himself drawn to politics. With his father, he had been involved in supporting a local parliamentary candidate, and he almost found himself put up for Parliament in 1762, at his father's wish. When he visited Paris, Gibbon found the English fêted as the champions of universal liberty. It is likely then that in conceiving the Swiss history Gibbon was making a bid for the attention, home and abroad, which the popular theme of liberty would have commanded him. It was a miscalculation.

If Gibbon truly set his heart upon being the historian of the decline and fall of Rome as early as October 1764, there was a frustrating delay before he devoted his attention exclusively to the project.

There was not merely the interlude of the history of the Swiss. With Deyverdun in England, Gibbon spent a considerable portion of 1768 and 1769 co-authoring with him two volumes – the only two volumes! – of the *Mémoires littéraires de la Grande Bretagne*, a French-language literary and scholarly review which they founded. It may not have been time lost, for it gave Gibbon practice in penning pithy comments on other authors, and thus training his literary voice. Characteristically, Gibbon the reviewer was a stickler for accuracy. In the account of James Boswell's *History of Corsica*, the author is advised to go and read Muratori and get his facts straight.

They were frustrating years. But Gibbon, kicking his heels during the late 1760s, steadily equipped himself for his larger project. He wrote, for instance, a series of historical essays. One, 'Du gouvernement féodal', surveyed the feudal system, rather unconventionally emphasizing its barbarian roots more than its potential for supporting sophisticated government. Another, 'Sur la monarchie des Mèdes', was an extended exercise in chronological criticism, which also presented 'various reasons for the decline and fall of the Median empire'. Gibbon deployed the learning of the old chronologers Marsham and Scaliger in the manner of the French *philosophe* Fréret, seeking above all to sift 'truly historical sources' from 'fabulous traditions', 'fictions' and gross blunders: all seemed dark, for the Persian writings had no 'geography, no chronology, [but only] paladins, genies, fairies and monsters'. Already Gibbon was chewing over some of the ideas which later preoccupied him, such as the capacity of custom and civilization to survive the ravages of time. Refuting the notion that the land of the Medes reverted to anarchy after its revolt against the Assyrians, Gibbon responds:

> a vast country ... could never return to the state of nature after having for more than five centuries borne the yoke of laws. Revolutions change the political contract but they never break the bonds of the social contract. The first is founded only on fear or prejudice. The habits and interests of all assure the eternal duration of the second.

Probably around 1770 Gibbon also wrote 'Outlines of the History of the World 800–1500', a bare annalistic account, organized century by century, perhaps as a draft sketch for the later volumes of the *Decline and Fall*.

The same period of Gibbon's life also saw two further sorties into scholarly warfare. The *Critical Observations on the Sixth Book of the Aeneid* (1770) was Gibbon's first English publication, albeit anonymous. In it he attacked the patriarch of English sacred historians, Bishop Warburton, author of the immensely influential *Divine Legation of Moses*. Warburton had contended that Virgil's account of Aeneas descending into the underworld was an allegory of the hero as lawgiver undergoing initiation into the Eleusinian mysteries. Gibbon resented what he saw as Warburton's appropriation of great poetry in aid of a partisan Christian polemic. He also thought the Bishop simply wrong, and, as recommended in his *Essai*, set out to show how true erudition refuted Warburton's idiosyncrasies. The pamphlet shows Gibbon already eager to joust against the heavy-weight supporters of the old sacred history, though at this stage taking the indirect approach, choosing Virgil not the Bible as the tiltground.

Gibbon attempted something a shade more daring in 1772 by taking issue, in a lengthy private letter, also anonymous, with the eminent religious scholar and *protégé* of Warburton, Richard Hurd, on that pivotal question for any sacred history, the prophecies of the Book of Daniel. Gibbon contended in true *philosophe* fashion that so literally had the 'prophecies' been fulfilled that common sense required us to believe that the writings themselves were forgeries, compiled *after* the events they pretended to predict.

Gibbon's onslaughts on Warburton and Hurd can be read as dummy-runs for the *Decline and Fall*'s far more comprehensive demolition of the pillars of conventional sacred history. They helped him let off steam during the unsettled time surrounding the deterioration of his father, whose declining health and deepening financial crises may have robbed Gibbon of the settled concentration required for serious study and writing.

For Gibbon that death in 1770 provided an almost tangible release. Between then and 1776 he ceased to be a man of preparation, apprenticeship and false starts hovering uneasily between critic and reviewer, scholar and gentleman, antiquarian and philosopher, *homme de lettres* and historian:

I more seriously undertook [from 1768 onwards] to methodize the form, and to collect the substance of my Roman decay, of whose limits and extent I had yet a very inadequate notion. The

Classics as low as Tacitus, the younger Pliny and Juvenal were my old and familiar companions: I insensibly plunged into the Ocean of the Augustan history, and in the descending series I investigated, with my pen almost always in my hand, the original records, both Greek and Latin, from Dion Cassius to Ammianus Marcellinus, from the reign of Trajan to the last age of the western Caesars.

Once installed in his smart new Bentinck Street house in 1772, with six servants, a pet parrot and a Pomeranian named Bath, there was no way of stopping him. 'No sooner was I settled in my house and library than I undertook the composition of the first Volume of my history.'

Gibbon had been an ugly duckling, and he indeed turned into a very singular young man, fat, fastidious and foppish. He had decided in his early twenties that he would never marry. He aimed to keep one foot in Europe and one in England: he wanted to study but also to taste the pleasures of company, he envied the scholars but also sought acceptance as a perfect gentleman. He aspired to the Horatian ideal, though he could also be satirized as 'Mr Chubby Chub', full of fussy bachelor mannerisms, the vain man hated by Boswell.

But the ugly duckling, misfit, even freak, unsure of his future, had also by the mid-1770s emerged as an intellectual swan, assured, confident and with a clear vision of his mission. The transformation was quite staggering. Psychologically we might attribute it to the death of his father: the cramps on his style were now removed. Now the master of his own life, Gibbon set himself up in London, moved in better circles, became a Member of Parliament in 1774, and through that experience grew immediately familiar with the noblest oratory in the English language.

Intellectually speaking, it is much harder to document what produced the ripeness of the early '70s. The days of copious journals cajoling himself into action, of lengthy commonplace books and fragmentary essays, are over. The youthful Gibbon had left plenty of trails to reveal his intellectual life, as he drove himself to become a scholar. Once he had succeeded, the trails end, though it seems that it was not until 1772 or '73 that Gibbon began composing his text in earnest. He tells us he wrote three drafts of the first chapter of the *Decline and Fall*, and two each of the next two, but thereafter the first draft was the final; as the sheets were written – he composed them a paragraph at a time in his head before committing a word to paper – so they were sent to the printer. How much, or how little, Gibbon tinkered, we don't know, because the final manuscript has not survived.

For many years Gibbon had been uncertain of the future; from 1776 he was sure of the past – becoming, to a quite remarkable degree, his own book. From then till the late 1780s, when he completed the *Decline and Fall*, we know oddly little of him, beside his small number of entertaining letters and those opinions which, framed in public form, made up his 'great work'. He had become 'the historian of the Roman Empire'.

3 The *Decline and Fall*

The first volume of the *Decline and Fall* appeared in 1776, when Gibbon was nearing forty. He had laboured long to produce a work worthy of the world, his labour and his hopes. In the end it was history which met these criteria. History seemed a particularly noble genre of learning and artistry: 'The subject of history is Man,' he commented, sententiously conveying the gist of the philosophy of Antiquity, the Renaissance and the Enlightenment; and, by implication, demarcating his enterprise from mainline sacred history. It also, Gibbon recognized, captured the attention: 'history is the most popular species of writing,' he reflected, 'since it can adapt itself to the highest or lowest capacity.' And this was so, ultimately, because it touched on matters closest to our hearts, questions of our own origins and destiny:

> A lively desire of knowing and recording our ancestors [Gibbon is here writing literally of genealogy, but his comment applies more widely] so generally prevails that it must depend on the influence of some common principle in the minds of men.

In the last resort, this 'principle' was the hope of a vicarious extension of our own existence, indeed a kind of immortality:

> Our imagination is always active to enlarge the narrow circle in which Nature has confined us. Fifty or an hundred years may be alotted to an individual; but we stretch forwards beyond death with such hopes as Religion and Philosophy will suggest, and we fill up the silent vacancy that precedes our birth by associating ourselves to the authors of our existence.

What form and method of history-writing, however, would be worthy of this elevated vision? Gibbon saw himself beset by object lessons in how not to do it. On the one hand, there was the cavalier approach to scholarship he believed was all too often displayed by the *philosophes*. Many of them felt a certain contempt for the past in any case. For Voltaire, who had proclaimed '*il faut écrire l'histoire en philosophe*,' history was essentially a tale of darkness and ignorance, terror and error. Only three previous epochs were worthy of the historian: ancient Greece, ancient Rome and the Italian Renaissance. The past was a dream from which we were trying to awake.

For that reason, Voltaire, as Gibbon saw it, approached history not primarily as the object of painstaking study in its own right, but with polemical intent. Gibbon treated him, somewhat unfairly, as rather a superficial historian, one who cast 'a keen and lively glance over the surface of history'. Moreover, the *philosophes* (Gibbon feared) had been over-impressed by the modish Pyrrhonism of the seventeenth century, especially that of Bayle. Faced with conflicts and gaps in the data they tended to despair of achieving any reliable historical knowledge, especially respecting the remote past (Samuel Johnson, it will be remembered, took a similar line on the limits of historical knowledge). History, judged Voltaire, is just a pack of tricks we play on the dead; Rousseau, approaching the early history of mankind in his *Discourse on the Origin of Inequality*, asserted cavalierly: '*commençons par écarter toutes les faits.*' Similarly Gibbon remarks of another French writer that he 'quotes nobody, according to the latest fashion of the French writers' (*plus ça change*).

Nevertheless, Gibbon learned much – far more than he admitted – from history as practised by the *philosophes*. They had pioneered a secular approach, they concentrated upon the epochs of man's greatest cultural achievements, they transcended traditional Eurocentrism, they looked for natural causes, and, not least, they championed social history, what Voltaire called *l'histoire des moeurs*, which focused not on courts and battlefields but upon the fabric of common experience. Even so Gibbon judged that the history the *philosophes* wrote too often played fast and loose with the truth. Not much was sacred to Gibbon, but facts certainly were. He was riled by Voltaire's lack of precision, and felt relieved when Voltaire actually turned to contemporary history, which he knew well enough, in his history of the reign of Louis XIV (1751):

I believe that Voltaire had for this work an advantage he has seldom enjoyed. When he treats of a distant period, he is not a man to turn over musty monkish writers to instruct himself.

He follows some compilation, varnishes it over with the magic of his style, and produces a most agreeable, superficial, inaccurate performance.

Gibbon of all people was no despiser of style, but he could not accept that verbal magic could deputize for learning. (Voltaire would have claimed to have been wearing his learning lightly.) And Voltaire's learning was apparently all too often deficient, as it amused Gibbon to point out often enough in his dynamite footnote rebukes: 'M. de Voltaire, unsupported by either fact or probability, has generously bestowed the Canary Islands on the Roman Empire,' runs one. At bottom, Voltaire could not command trust as a scholar; since he rarely gave footnote references, it was impossible to tell on what evidence his judgments were based.

Gibbon was also perturbed by some of the more pretentious historical theorizing of his day. It had become the Paris fashion to slight erudition about the past as a load of lumber; and so histories tended to be painted with a broad brush, thus leading to gross simplifications. In the second half of the century bold canvases of this kind came into their own in the philosophical histories produced in France by Turgot, Volney and Condorcet, in Scotland by Adam Ferguson and John Millar, and in Germany by Herder. Such works aimed to capture a philosophical grasp of the stages of human development, demoting the precise details of the factual record as subordinate, even expendable. But it was a genre already well established by the time Gibbon defined his historical priorities. Attempts to lay bare the anatomy and physiology of society, showing the secret springs of socio-political growth and decay, were already popular. In 1695 Vertot had produced his *Revolutions of Sweden*, in 1719 his *Revolutions of the Roman Republic*; in 1749 there appeared Lévesque de Burigny's *Revolutions of the Empire of Constantinople*, and in 1750 the first volume of Marigny's *Revolutions of the Empire of the Arabs*. Such dramatic visions of human destiny were continued later in the century by the Abbé Mably and above all by Volney, most famously in his popular *Ruins of Empire*, which embroidered a dazzling historical panorama with popular moralizing and a prophetic vision: too bad that it was almost devoid of facts.

But the idea of a pathology of political society had above all been given intellectual cachet by the redoubtable Montesquieu, with whose ghost Gibbon engaged in a lasting dialogue. Gibbon had great respect for Montesquieu and paid him the compliment of echoing his style in the *Essai*. Unlike Voltaire, Montesquieu's seriousness could not be gainsaid. He possessed an enviable capacity for discerning patterns, for plumbing deeper meaning, in the most miscellaneous data. Most obviously in his *Considérations sur les causes de la grandeur des Romaines et de leur Décadence* (1734) but also in his *magnum opus, De l'esprit des lois* (1748), he had sought to grasp society as a totality, revealing the interplay of environment and people, social structure and social action, consciousness, institutions and political power. In the *Spirit of the Laws* (1748) in particular, he had advanced a taxonomy of political societies, categorized not primarily according to their constitutional form (rule by the one, the few or the many) but according to the power relations permeating the very fibres of society. The distinct 'natures' of different polities (republican, monarchical and despotic) were characterized by their own 'spirits'. Thus the lifeblood of the republican mode lay not only in its constitution but also in a spirit of 'virtue', by which Montesquieu meant an ardent civil patriotism and the flame of liberty. The 'principle' of monarchy was 'honour', or an overriding attachment to rank and hierarchy. Despotism was a degenerate type of rule, whose governing principle was 'fear'. In Montesquieu's view, each type of social organism followed, by a kind of natural law, autonomous pathways of growth and development, under the pressure of natural forces such as terrain and climate; for instance, hot climes tended to enervation, and thus disposed regimes towards slavery and despotism.

Montesquieu believed that in these analytical principles, grounded in the nature of things, lay the key to Rome's decline. As the Roman republic grew bigger and wealthier, it necessarily lost the spirit of 'virtue' which had spurred its growth: 'The corruption of each government generally begins with that of its principles.' As the successful nation changed its nature, its principle was lost: 'the greatness of the republic was fatal to its republican government.' Everything hung together:

> Here [explained Montesquieu] is the history of the Romans in a nutshell: they conquered all people by their maxims; but when they achieved this, their republic could not subsist: it was necessary to change the government; and maxims contrary to the former, employed in the new government, were the cause of their downfall.

But was this insight – it has been called 'sociological' – into the body politic illuminating or just a slick verbal formula? Though impressed by Montesquieu, Gibbon inclined towards the latter view. As several of his critical footnotes indicate, Gibbon worried that Montesquieu's generalizations were not always congruent with fact. Indeed, he seemed occasionally guilty of extraordinarily damning ignorance or error, as when he wrote that, during the reign of Valens, the Goths crossed the Danube, ravaged everything, *and then retreated back across the river*. This was a capital blunder:

> The President Montesquieu seems ignorant that the Goths, after the defeat of Valens, *never* abandoned the Roman territory. ... The error is inexcusable, since it disguises the principal and immediate cause of the fall of the Western Empire of Rome.

Gibbon was not nitpicking over a silly slip; fundamental issues of interpretation and explanation were at stake. For the 'sociologist' Montesquieu, the essence of Rome's collapse lay in inordinate growth and consequent inner decay, hence precisely what the Goths were doing and where they were did not matter too much; for Gibbon the historian, by contrast, Rome fell because of a concatenation of many circumstances, some internal, some external. The Goths were not just optional extras in the Roman tragedy.

In other words, Montesquieu, focusing upon the relations between structure, function and ideology, believed that there were 'ideal types' of societies, conforming to general patterns of growth and decay. Rome exemplified such a typology. Gibbon, however, places little faith in the theorists' desire to abstract and generalize in this way. Not least, though appreciating Montesquieu's insights into impersonal causes and unintended consequences, Gibbon was sceptical about ascribing historical effects to such natural factors as climate. In contrast to Montesquieu's attempt to explain history through 'sociology', or Hume's emphasis upon psychological causes (invoking the 'universal principles of human nature'), Gibbon explained history through history, by a detailed narration of the total interconnectedness of events.

Thus Gibbon had deep reservations about philosophizing about history, especially if it involved system-building and espoused a sloppiness over facts which demeaned the enterprise. Always severe on the culpable blunders of his historical brethren, his notebooks abound with quips such as 'Leti is a most agreable historian, a little more regard to truth and exactness would have made him an instructive one.' Accuracy was of cardinal importance. Moreover, for Gibbon much of the joy of history lay in the sheer profusion of knowledge about the past which the scholar could communicate. The reader of the *Decline and Fall* is regaled not only with an immensely rich tapestry of the political and military affairs of Rome, not just with learned dissertations upon subjects which a less erudite scholar might safely have omitted, such as the intricacies of the Justinian Code or the theological niceties of the early Church, but with a whole range of other information on top, not omitting an account of the Russian sturgeon fisheries ('the endless exportation of salt fish and caviar is annually renewed by the enormous sturgeons that are caught at the mouth of the Don or Tanais'), or a revelation about the mysteries of Bologna sausages ('said to be made of ass-flesh').

Gibbon paraded his learning not just to show off, but to bolster confidence in his authority to speak. The *Decline and Fall* contains some 8362 references. Gibbon, who confessed to cultivating an 'instinctive love and accurate study of ancient Geography and Chronology', piqued himself on getting such minute details right as distances, battle-sites, town plans, army route-marches, weights and measures, and the like – while avoiding mere antiquarian love of facts for their own sake.

Precision was the indispensable foundation for history:

> Diligence and accuracy are the only merits which an historical writer may ascribe to himself; if any merit indeed can be assumed from the performance of an indispensable duty, I may therefore be allowed to say, that I have carefully examined all the original materials that could illustrate the subject which I had undertaken to treat.

Thus Gibbon in the 'Advertisement' to his footnotes – which he originally placed as endnotes but (so he says) by 'public importunity' moved them, from the third volume onwards, to the foot of the page: one suspects the reason is that they are the most entertaining, malicious and often salacious footnotes in history.

Was Gibbon's bold claim to have 'examined all the original materials' true? No, if by that we include the thousands of manuscripts in European archives and libraries which Gibbon never consulted. Yes, if he means that he had not cobbled together a compilation out of other compilations, but had personally scrutinized all the best printed editions of the primary sources, and a mountain of other books besides – travel literature, geographies, topographies, biographies and so forth. Thus, for example, the *Decline and Fall* contains some 120 references to the *Memoirs of the Academy of Inscriptions* and about the same number to 'that ingenious geographer', d'Anville, an 'excellent writer' whom Gibbon habitually compliments on his infallibility with names and locations (he caught him out only once); he shows intimate knowledge of many other travel writers besides, above all the Frenchmen Tavernier, Thévenot and Tournefort.

And Gibbon spoke no more than truth when he claimed that he had used his sources 'carefully'. Searching for an Achilles heel in his scholarship, his theological detractors alleged that his learning was second-hand; but it was their own ignorance which they exposed in their unsubstantiated claims, as Gibbon magisterially demonstrated in his *Vindication* (1779). As the great critic Richard Porson remarked, in squashing one of these insects, 'his industry is indefatigable, his accuracy scrupulous; his reading, which indeed is sometimes ostentatiously displayed, immense; his attention always awake; his memory retentive.' Porson's judgment merely echoed the compliment paid Gibbon by his brother historian William Robertson, who confessed, 'Before you began your historic career, I used to pride myself in being at least the most industrious historian of the age; but now, alas! I can pretend no longer even to that praise.' Robertson tried checking Gibbon's footnotes:

> I have traced Mr Gibbon in many of his quotations (for experience has taught me to suspect the accuracy of my brother penmen), and I find that he refers to no passage but what he has seen with his own eyes.

Gibbon passed Robertson's test magnificently – 'He possesses that industry of research without which no man deserves the name of an historian.' David Hume likewise complimented Gibbon on 'the depth of your matter [and] the extensiveness of your learning'. Bury's great early twentieth-century edition showed where nineteenth-century scholarship had rendered Gibbon obsolete; but the tally of blunders, slips of the pen or examples of culpable ignorance which Bury detected is quite amazingly small.

As mentioned earlier, Gibbon did not originate new historical techniques, nor did he fully exploit some of the auxiliary historical sciences being developed in his own day, such as diplomatics or palaeography ('I studied the theory, without attaining the practice of the art'). He made good use, however, of the major works of numismatics, Ezechiel Spanheim's *De Usu Numismaticus* (1671) and Charles Patin's *Histoire des médailles* (1665). His bedrock was written evidence. That he had made no breakthroughs in source criticism was pointed out early on by a reviewer in the *Göttingische gelehrte Anzeiger* of 1788, though, being unable to read German, Gibbon never saw it. Gibbon instead rested content with a general rule-of-thumb with sources: the older the better; but he knew them like the back of his hand, had an eagle eye for bias, was adroit at resolving contradictions between rival authorities, and above all thought hard about what could be extracted from the testimonies of the past.

What qualified Gibbon to be the 'historian of the Roman Empire' was the simple fact that he knew more about it, and had thought more deeply about it, than anybody else. This had demanded prodigious labours, regular work habits and an exemplary capacity for taking pains. But in some ways his task had been made relatively easy. The scholarly drive of the seventeenth century had led to the publication of an immense body of historical documents – laws and bulls, charters and codes, annals and chronicles – and the compilation of histories incorporating them. This was the product, as we have seen, of the religious warfare arising out of the Reformation and Counter-Reformation, of the humanist fascination with preserving, purifying and printing sources, and of the antiquarian movement.

Religious partisanship had in some cases sabotaged scholarly standards. Gibbon's footnotes mark his exasperation with those who put advocacy before truth, such as the biased Protestant *Magdeburg Centuries* (a sixteenth-century chronicle of Church history), or the work of Cardinal Baronius, who in his *Annales Ecclesiastici* had 'sunk to the lowest degree of credulity which was compatible with learning' (yet, because he had combed the Vatican library, Baronius' work, though well over a century old, was still indispensable). Nonetheless many of these religious scholars had in fact been successful in surmounting partisanship, so strong was their conviction that truth was great and would prevail.

The fruits of numerous Catholic scholarly enterprises were readily available to Gibbon. He made great use of the works of the Bollandist scholars, above all their *Acta Sanctorum*, begun in 1607, a vast compilation of the lives of the saints; it was, said Gibbon, 'an undertaking which, through the medium of fable and superstition, communicates much historical and philosophical instruction'. He also drew heavily upon such monastic scholars as Dom Bouquet, Dom Rivet and above all Montfaucon, the monk of St Maur, a scholar with a vast European-wide correspondence network. Montfaucon had edited the Church Fathers and pioneered interpretation of the physical remains of Antiquity, publishing his findings in the ten volumes of his *L'Antiquité expliquée* (1719).

Pride of place, however, went to the Jansenist Le Nain de Tillemont (1637–98). This seventeenth-century *érudit* covered the first six centuries of the Christian era in his two monumental works, the sixteen-volume *Mémoires pour servir à l'histoire ecclésiastique des six premiers siècles* (1693–1712) and his *Histoire des Empereurs* (1690–1738). His project was to integrate all the major sources for the early history of the empire and the Church, paying scrupulous attention to questions of dating, attribution and authenticity, and resolving conflicts between them. To modern eyes, the Jansenist was occasionally at odds with the scholar. For one thing, Tillemont regarded the records of the Church Fathers as *ipso facto* more reliable than those of the late pagan Romans. Sometimes bemused after his scans of the past for signs of divine justice left him puzzled ('You have visibly destroyed Nero, Domitian and others. But did You destroy Trajan and Marcus in the same way? ... They persecuted Your servants'), Tillemont could betray an enchanting perversity towards his sources, rejoicing on one occasion that Providence in its wisdom had permitted all records of the early years of Diocletian, that persecutor of the Christians, to perish:

> God, who opposes the proud, has thus permitted the loss of these works in order to confound the vanity of the insolent usurper of His divine name; the usurper of the adoration which is due only to God Himself.

But a Jansenist humility instilled in him an unremitting accuracy to which Gibbon pays wryly condescending compliments throughout the *Decline and Fall*. In his twenty-two volumes, Gibbon notes, Tillemont may be 'bigotted', but his bigotry is 'overbalanced by the merits of erudition, diligence, veracity and scrupulous minuteness'; as a consequence, he is 'the patient and sure-footed mule of the Alps' who 'may be trusted in the most slippery paths'.

If 'that incomparable guide' led Gibbon safely into the Dark Ages, he was led out by Ludovico Muratori (1672–1751), librarian of the Ambrosiana at Milan, and his volumes of *Scriptores* – a vast collection of the works of historians of Italy from the sixth century through to the Renaissance. Gibbon's appreciation of Muratori was greatly enhanced by his other work, the *Antiquitates Italiae*, a collection of assorted documents and critical dissertations on the life of medieval Italy. Muratori received fair tribute:

> through the darkness of the middle ages I explored my way in the Annals and Antiquities of Italy of the learned Muratori; and diligently compared them with the parallel or transverse lines of Sigonius and Maffei, Baronius and Pagi, till I almost grasped the ruins of Rome in the fourteenth century.

Many other works helped Gibbon on his way. Godefroy's writings enabled him to make shrewd use of the Theodosian Code, a collection of laws issued in 439, as a source for the social history of the empire in decay. Godefroy's interpretation of the Code offered sure evidence of the growing burden of taxation and of land falling out of cultivation; Gibbon picked up these insights. He also leaned on the history of Manicheanism by Beausobre ('a Protestant but a philosopher'), which was 'a treasure of ancient philosophy and theology' – yet another pious work which Gibbon gutted for his own purposes, going 'beyond the holy circle of the author'. The German Lutheran Mosheim ('full, rational, correct, moderate') was a major source for Church history; Mosheim was remarkable for surveying it not as the pattern of Providence but as a civil institution. For Italy, Gibbon utilized the *Civil History of Naples* (1723) by Pietro Giannone – not just a chronicle but a pioneer essay in *storia civile*, the critical, philosophical history of civil society; for France he drew upon the eleven-volume *Recueil des historiens des Gaules et de la France*, edited by Dom Bouquet and his fellow Benedictines (a work, Gibbon noted, pondering the sad lack of scholarly early histories of England, which might 'provoke our emulation').

Where Gibbon's sources are strong, his history is at its best. Where the printed materials were meagre or poorly edited, as in the case of Byzantine history or early Islam, his own scholarship suffers. Gibbon is, however, typically willing to confess his own limitations. He writes his history of Islam, he admits, without knowing Arabic: it was not wholly his fault, he had wanted to study it at Oxford. Sources for the life of the Prophet are particularly lacking, and, dealing with Mohammed, he largely leaves it to the reader to make up his own mind (a fanatic or a fraud?). Interestingly, it was in his Arabic sections that Gibbon found the most English secondary sources to draw upon; he makes fifty-five references to the seventeenth-century Oxford scholar Edward Pococke.

Deep learning and accuracy were essential for history. So was an attention to detail, not just for its own sake, but for the meaning which could be derived from it: it was vital to place the individual deed or character within the wider framework of events and interpretations. In other words, true history was not simply one damned fact or thing after another. Out of humility, severity and austerity, and perhaps for want of skill, the Jansenist Tillemont (and many of the other antiquarians noted in Chapter 1) had reduced history to drab chronicle. Take a sample of Tillemont's notion of narrative. He is here using Eusebius as a source for Diocletian's defeat of the usurper Achilleus in 296:

> He says that the cities of Coptos and Busiris, in the Thebaid, having revolted against the Romans, were destroyed and razed almost to their foundations. The Greek text attributes this action to Diocletian and Hercules [Maximian], which account also appears in Theophanes: ... Father Petau fixes the ruin of Coptos and Busiris in AD 298. Zonoras places it before the revolt of Achilleus, and reports that only Diocletian participated in the action.

Accurate and painstaking this undoubtedly is. History (according to Gibbon's notion, and ours) it is not. For one thing it is unreadable; and – as I shall discuss below – Gibbon believed that a fine style was more than a frill, as it formed the medium for meaning. For another, it is flatly and banally descriptive: it tells us nothing about character and motive. Mosheim was much the same. He organized his history strictly by years; thus chronology took precedence over actions and events. History recorded as annals begins to establish what happened. It says nothing, however, about causes, significance, implications, consequences: it affords no interpretation. That, for Gibbon, was the crucial distinction between mere chronicle and real or, as he would say, *philosophic* history.

Gibbon distrusted the intellectual pretensions and polemical goals of the *philosophes*, but he was deeply committed to interpreting history through philosophical eyes. As he put it in an apothegm, 'if philosophers are not always historians, it were, at any rate, to be wished that historians were always philosophers.' What did he mean by this? In part, a faith in the meaningfulness of history: history was not just a tale full of sound and fury. And in part it entailed, following Voltaire and Montesquieu, a commitment to explanations within the realm of the human and the natural, rather than the divine. Thus Gibbon criticized ancient Baronius as 'more inclined to seek the cause of great events in heaven than on earth'. History was not a cosmic puppet theatre, in which Providence pulled the strings.

But what else did Gibbon mean by historians being philosophers? Certainly not a commitment to some metaphysical scheme, which he always deplored as nonsense on stilts. Rather the play of a cast of mind, a particular relational mode of thought. This notion of thinking 'philosophically' was clearly expressed in his *Essai*:

> The philosophic spirit consists in being able to go back to basic ideas; to perceive and to bring together first principles. The view of the philosopher is exact, but at the same time extensive. Placed on a height, he comprehends a great expanse of territory, of which he forms a clear and unique image, while other minds just as exact, but more confined, discern only a part of the expanse.

What then did being philosophical actually entail in the quotidian practice of history? For one thing, it meant having the mind to avoid taking sources at face value, but instead to cross-question them intelligently. Like the earlier Pyrrhonists, Gibbon recognized that sources often contradicted each other. Obviously he knew this was often due either to unconscious bias or to quite specific propaganda purposes: thus the court chroniclers of the Byzantine dynasties produced not histories of emperors but panegyric, and the monkish chroniclers wrote hagiography. Unlike the Pyrrhonists, however, Gibbon, when faced with such distortions, was disinclined to pause and lament the frailty of man, the darkness of the past or the vanity of learning. Rather 'our immortal reason' had to apply itself more aggressively to the materials. Gibbon had learned the scholars' lesson, that the correlation of different accounts could establish scales of reliability. But he also believed that sources could be systematically cross-examined, their biases and blind spots identified, and truth winnowed from fiction, or at least a probability gradient established. Gibbon draws attention, for example, to an account of the Emperor Severus Alexander in the *Historia Augusta*, a highly unsatisfactory compilation of late Roman history but the only one to survive. Carefully examined, it can be seen to derive from the *Cyropaedia* of Xenophon, that blueprint for the education of a perfect prince. Thus it is perhaps of dubious value for the life of Severus himself. Yet that need be no cause for despair. For what it clearly did establish was the *mirror* for princely behaviour dear to the late empire; it also suggested the growing servility of authors, their retreat into panegyric. Gibbon's footnotes expose the limitations of his sources, but they go beyond this negative achievement, showing that shrewd reading can disclose nuggets of information amidst the dross of error, ignorance and prejudice.

The philosophical historian must thus train a rational intelligence upon his sources and engage in a fruitful dialogue with them. Gibbon was the master of this art; as G. M. Young so deftly put it, he 'applied the mind of the eighteenth century to the learning of the seventeenth'. But, as Gibbon highlighted in his own account of being 'philosophical', the historian must also display a mind of uncommon breadth, with an angle of vision wide enough to escape from chauvinism and ethnocentricity. He must distance himself from prejudice, integrate the particular and the general, and take a 'clear and comprehensive view of the revolutions of society; [so he is] capable of discovering the nice and secret springs of action, which impel, in the same uniform direction, the blind and capricious passions of a multitude of individuals.' The temper of Gibbon, and of other Augustans such as Johnson, was one forever comparing and relating, in order to cut things down to size or see things in perspective. Thus, for instance, the nearest Gibbon gets to painting a sympathetic portrait in the *Decline and Fall* is the Emperor Julian (known to Christian historians as the 'Apostate'). Gibbon indulges himself by lavishing three chapters upon his brief reign. Julian is praised as the model of a martial emperor in the old mould. But Gibbon is careful to weigh his greatness against his shortcomings. Julian showed a fatal weakness for metaphysics and Platonic speculation, and his fanatical attachment to the old paganism ironically leaves him as the mirror image of the Christian fanatics he and Gibbon despised. Likewise, Gibbon admired ancient Rome; but as a historian he avoids civic hagiography or nostalgia. The Romans thought their city was eternal, but the philosophical historian, from his Olympian height, shows time and affairs humbling their pride. Rome had left Carthage a heap of ruins. Now cattle were browsing, friars chanting and Gibbon musing, amidst the ruins of the Capitol itself. Philosophical reason must counter the passions of pride, prejudice and enthusiasm.

Above all, breadth for Gibbon meant taking a vast panorama. The *Decline and Fall* is not merely 'another damned thick square book' – as a good Augustan, Gibbon was well aware that vastness is rarely greatness. It is magisterial in its theme, scope and vision. It offers a fine-textured account of a millennium and a half of history: though formally commencing in AD 180, it actually tracks back to the inception of the empire, and traces the destiny of Rome through to the sack of Constantinople in 1453. It ranges from the frozen wastes of Siberia to the cataracts of the Nile, from Korea to the Pillars of Hercules, and encompasses, within its asides, the New World of America and the antipodes of New Zealand.

Rome of course dominated the Augustan imagination. Scores of accounts of pagan Rome and Christian Rome – historical, artistic, architectural, topographical – had been published in seventeenth- and eighteenth-century England in works such as Basil Kennett's *Romae antiquae notitia* (1696), Joseph Addison's *A Letter from Italy* (1701) and George Keate's *Ancient and Modern Rome* (1755). Its decay had become a well-worn poetic topos, as in John Dyer's *The Ruins of Rome* (1740). William Wotton had already brought out his 'trailer' to Gibbon, *The History of Rome from the Death of Antoninus Pius to the Death of Severus Alexander* (1701), a work which anticipates the *Decline and Fall* in its periodization of collapse, though it amounts to little more than a series of factually sound biographies of the emperors. Gibbon's '*songe d'antiquité*', was thus by no means unusual: as he himself put it, 'Rome is familiar to the school-boy and the statesman.'

Within the compass of his seventy-one chapters Gibbon recounts 'the memorable series of revolutions, which, in the course of about thirteen centuries, gradually undermined, and at length destroyed, the solid fabric of Roman greatness'. It is the story of two empires, the Western and the Eastern. It treats the rise of Christianity, seen both as an institution and as a creed, and deals copiously with other faiths too, especially Zoroastrianism and the rise of Islam. And it looks beyond the Roman borders, particularly to the barbarians but also to civilizations which were its traditional enemies, such as the Persian. Gibbon unfolds the barbarian paradox: they conquered the great empire and the eternal city, but they were in turn conquered by the great idea – civilization – which they had fought. Gibbon has an assured feel for the pulse of history, the tides of national destiny, the ironies of ambition, the illusions of achievement, the concatenating cycles of rival empires advancing and receding over plains and deserts and seas. Astounding heroes are brought to life: Constantine, Julian, Boethius, Alaric and Attila, Genghis Khan and Tamerlaine. Gibbon marshals and organizes his massive documentation so that the stupendous architecture of the whole, far from being masked by detail, is revealed all the more clearly.

Gibbon's classical training and Augustan temper had left him sensitive to the patterns of order and repetition, to the parallels, symmetries and dramatic ironies which inform the sweep of Roman history. His work opens on a pivotal point, the death of Marcus Aurelius in AD 180. Was not this the very pinnacle of Roman greatness, the most felicitous moment in world history?

> In the second century of the Christian aera, the empire of Rome comprehended the fairest part of the earth, and the most civilized portion of mankind.

The first three chapters take stock of the state of Roman greatness under the early empire; Gibbon reviews its geographical extent, province by province, surveying the wealth, agriculture and peoples of each; the military might of the empire is paraded before us; we see the prosperity of the city itself, its culture, religion and arts, as well as the teeming urban life of Rome; and we are treated to a historical analysis of imperial government, from Augustus himself, through the troubled times of the mid-first century AD, to the happy century of the five good rulers, Nerva, Trajan, Hadrian, Antoninus Pius and Marcus Aurelius. But in the course of these chapters the fatal flaw of the system is revealed. The empire is the greatest the world had seen. Imperial rule secured for the Romans at least one happy century of peace and prosperity. But the empire was weak in its strength. For it depended ultimately, indeed increasingly, upon its emperor. Even the philosopher Marcus Aurelius proved the wisest fool in the empire, since he made his empire over to his inept, irresponsible adoptive son Commodus.

The fourth chapter hence launches into the first visible signs of decline. The troubled reign of Commodus results in his murder. Pertinax succeeds, but in turn is himself slain by the praetorian guards, who, once the tools of the emperors, threaten to become their masters. Murder breeds murder and worse. On Pertinax's assassination by the praetorians the empire is put up for auction; rival commanders battle for the prize; Septimus Severus eventually wins, thanks solely to the army (for which reason 'posterity justly considered him the principal author of the decline of the Roman Empire'). Subsequent chapters show the rot worsening. The first half of the third century, stained by the 'tyranny of Caracalla' and the 'follies of Elagabalus', is a political and military tragicomedy: no fewer than twenty men claimed the purple within fifty years; emperor after emperor is assassinated.

These divisions within the empire are accompanied offstage, as Gibbon skilfully shows, interweaving narratives of internal and external events, by the rise of the Goths (chapter 9), who irrupt into the empire itself in 269, only to be defeated by Claudian and Aurelian. The latter, in turn, is assassinated while campaigning in the East. Half a century of military turmoil and Gothic invasion is settled by the imposition of embryonic imperial despotism by Diocletian, whose work is then brought to maturity by Constantine. He transfers the seat of empire to the East, to the new Rome on the Bosphorus, and converts to the Christianity whose rise is explained in the controversial final chapters (15 and 16) of the first volume.

The plot, both majestic and subtly ambivalent, has been set in motion. The *pax romana* has already been assailed by internal discord and by barbarian attack from without. To meet these challenges, emperors were driven to administrative centralization and religious absolutism. What had been the glories of the serene empire – its peace, liberty and tolerance – were thus sacrificed to ensure its very continuation. The 'slow poisons' of these therapeutic measures will be seen tainting the subsequent centuries.

It is needless to offer here a guided tour of Gibbon's grand vision, chapter by chapter. The second volume, which together with the third appeared in 1781, opens at the fourth century AD with an analysis of Constantine's political regime and the implications of the imperial capital's transfer to the East. The volume is dominated by two contrasting portraits, those of Constantine and Julian. Constantine is pictured as a fearful and insecure man resorting to craft and stealth, and bargaining away the future by his own expedient adoption of Christianity as the state religion. 'As he gradually advanced in the knowledge of truth, he proportionately declined in the practice of virtue.' What Gibbon means by this stroke of irony is that, as soon as he became a Christian, Constantine felt free to murder his wife and sons. Ever alert to *déjà vu*, Gibbon seizes on the chance to point the contrast between Augustus and Constantine:

> In the life of Augustus, we behold the tyrant of the republic converted, almost by imperceptible degrees, into the father of his country and of human kind. In that of Constantine, we may contemplate a hero, who had long inspired his subjects with love and his enemies with terror, degenerating into a cruel and dissolute monarch, corrupted by his fortune or raised by conquest above the necessity of dissimulation.

Julian then enters as a foil to Constantine. He is an emperor true to the old Rome, who aims to restore the original principles, not least its paganism. Yet the clock cannot be put back, and Julian too is flawed, for he has become fatally infected with the lures of the East, to which the empire had migrated.

After Julian's death, an empire painfully stretched between Rome and Constantinople is further menaced by East–West divisions, above all during the period of dual rule by Western and Eastern emperors; and, contradicting the Christian principles of passive obedience, the throne becomes sucked into a maelstrom whipped up by the gales of Christian heresy, schism and unorthodoxy. Meanwhile – and Gibbon slips in these chapters like a master dramatist unfolding multiple plots at once – the barbarians have been massing. Themselves driven west and south by the Huns, the Goths spill over into the empire (chapter 26). Rome is defeated at Adrianople in 378, and the victors ravage the provinces; Theodosius soon overcomes and temporarily pacifies them; but a turning-point has been reached, for the pacified Goths are now settled within the empire.

The beginning of the end of the old empire is thus now in sight, for the Western remnant has been left hopelessly exposed to barbarian attack. Much of Volume 3 surveys further revolts of the Goths, the descent of the Alemanni into Gaul and of the Visigoths into Spain, the independence of Britain, and Alaric's invasion of Italy. In 410 Rome itself falls. Shortly afterwards, the Huns invade, Rome is sacked once more by the Vandals (the 'Final Destruction'), and we attend the abdication in 476 of the last Roman emperor in the West. By a nice philological coincidence his name, Augustulus ('little Augustus'), combines the syllables of Romulus and Augustus; unfortunately he did not inherit their characters, being 'a youth recommended only by his beauty, [who] would be the least entitled to the notice of posterity, if his reign, which was marked by the extinction of the Roman Empire in the West, did not leave a memorable aera in the history of mankind'.

Cunningly, however, Gibbon does not terminate his volume or vision with the collapse of imperial government in the Western provinces. Rather he insinuates a new theme – of rebirth and growth – with the establishment of the sapling barbarian kingdoms in the West, the Vandals in Africa, Merovingians in Gaul and Visigoths in Spain. We see Odoacer crowned the first barbarian King of Italy (chapter 36): as Rome is barbarized, the Goths are Romanized. Once again, Augustus forms the yardstick:

> It had been the object of Augustus to conceal the introduction of monarchy; it was the policy of Theodoric to disguise the reign of a Barbarian. If his subjects were sometimes awakened from this pleasing vision of a Roman government, they derived more substantial comfort from the character of a Gothic prince who had penetration to discern, the firmness to pursue, his own and the public interest.

With characteristic Gibbonian perspicacity, two crafty rulers are paraded, each concealing his true intentions but in sharply contrasting ways; and with characteristic Gibbonian generosity – he was never a simple partisan of Rome – a tribute is paid to the realism of the Goth who replaced the Roman myth with something substantial. We then get a glimpse of the emergence of the Frankish state in France after Clovis, and finally – a signal, surely, for the home crowd to cheer – the Saxons in England. Might has fallen, but, as the ever subtle Gibbon is concerned to show, the barbarians themselves are becoming civilized and have taken over the trappings, at least, of Rome. The third volume – at the close of which Gibbon paused, unsure whether to continue – concludes with the famous 'General Observations on the Fall of the Roman Empire in the West' (chapter 38), which will be assessed below in Chapter 6.

The last three volumes offer an artful reprise, almost a mirror image, of the first three. Volume 1 had begun with Rome apparently at its majestic height, to be gradually undermined by discord within and enemies without, in 'the triumph of barbarism and religion'. Volume 4 opens with the Eastern empire at its apparent height, during the reign of Justinian in the early sixth century. Subsequent chapters then trace *its* downfall, the victim of comparable forces. The internal weakness of the Byzantine court is surveyed in chapter 48, which exposes the stupidity of five centuries of Eastern emperors within a hectic seventy pages – a chapter rightly said to be the weakest in the whole work, but one which through its almost comic compression conveys the relative passivity of the Byzantine emperors at the eye of the storm. The new barbarians of the East, the Goths of the desert, are the Arabs. Unlike the Goths, however, the Saracens are driven to invade and conquer by religion, and Gibbon charts the rise of Islam in the early chapters of the fifth volume. The dramatic irony is now complete. Internal theological dissension and the rise of barbarism have divided the old empire into rival Christian blocs, to be in turn assailed by the Arabs in a kind of holy war.

For Christendom itself, despite its zealous insistence upon monotheistic unity (one God, one faith), has fallen victim to schism and partition. Gibbon traces, through the fourth volume, the thorough clericalization of society, politics and government, and the soul-sapping disputes over competing heresies (the Nestorians, Monophysites, Copts, Armenians, etc.) which split the Church in the East and eventually led to the total schism of the Roman and the Eastern Churches. Further variations follow on the theme of the destructiveness of religion and barbarism. Crusades proclaimed against the Saracens eventually turn against the heart of the empire itself: and Constantinople is sacked by the Christian crusaders in 1214, before falling victim a second time, and decisively, to the Ottomans in 1453.

Yet, by a curious irony, Rome itself survives. Commentators have been puzzled that Gibbon did not devote more of his final volumes to the flourishing of Latin Christendom in the West, to Charlemagne, the Papacy, the German emperors and Rome's ghostly afterlife in the Holy Roman Empire. All these do indeed get scanty treatment, and Gibbon's own relative indifference to things medieval must have something to do with it. But, once again, this choice reflects his conception of the unity and integrity of his subject. He is not writing the history of the transformation of the ancient world, of the Middle Ages, or even of the rise of the West. It is Rome which is his song, and his last chapters eventually call us back to Rome itself, the embarkation point for this long voyage. The closing sections of the final volume revert to the medieval history of the city, the political popes and the Roman aristocracy, those senatorial spectres; and the attempts to found a new republic in the ashes of the old led by Arnold of Brescia and Rienzi ('hero and buffoon'), both of whom seem to Gibbon like pantomime parodies of ancient Romans. Rome cannot, however, be revived in reality, and so it must live on in the mind. Hence Gibbon ultimately turns to the cultural regeneration of the West, the revival of letters associated with Petrarch and his poetic coronation in Rome; and we end in pastoral mode amidst the ruins of ancient Rome itself, elegiacally wafting from Poggio's invocation of departed glories to Gibbon's own.

Précis can only impoverish the richness of the seventy-one chapters of Gibbon's Roman tragedy. But the stark simplifications which paraphrase demands may delineate the grand architecture of Gibbon's design which, like a classical building, reverberates with internal echoes, symmetries and parallels; themes appear, are restated and then repeat themselves, often distinguished or reversed. It is as though some Demiurge were shaping human history rather like snowflakes, producing endless variations upon a standard crystalline structure: indeed, Gibbon liked referring to God as the 'Great Author'.

In these days when even readability – to say nothing of literary aspirations – in a historian tends to excite suspicion, it would be easy to speak slightingly of the design of Gibbon's work as if it were just a trick of 'style', merely a knack of plotting; it has even been suggested that Gibbon sacrificed the integrity of his material to the exigencies of a good story, replete with dramatic climaxes and ironies. Indeed, he ensured that each volume ended at a point of dramatic suspense.

But such strictures would be misplaced. To denigrate 'style' in history would be, for one thing, to adopt a foolishly positivistic view of what history should aspire to (a scientific report), as well as to express a puritanical attitude towards writing itself. Gibbon certainly held no brief for style for style's sake, and often deplored its excesses, dismissing the late Roman historian Ammianus Marcellinus, for example, with the comment that he 'is so eloquent that he writes nonsense'. Yct Gibbon believed that expression was no mere literary adornment, but rather the index of thought: 'the style of an author should be the image of his mind.' Moreover, without 'art and study' there could be no style at all, for expression was not the free electricity of genius, but the perspicacious matching of words and images; language must embody meaning, and the right voice did not come without studious craftsmanship:

> many experiments were made before I could hit the middle tone between a dull Chronicle and a Rhetorical declamation; three times did I compose the first chapter, and twice the second and third, before I was tolerably satisfied with their effect.

In other words, Gibbon was not one of those historians who believed, like naive positivists, that historical truth emerges automatically once a critical mass of data is reached. Historical interpretations are made out of the play of mind upon matter, and thought is a function of language (being bilingual, Gibbon was in a favoured position to understand that). In Gibbon's view, the historian's free and critical spirit acts as the intellectual equivalent of Providence or Time. It makes, or remakes, history in the mind, after the fact, reading the true patterns out of, or perhaps into, the flux of events.

Gibbon's belief in the confluence of medium and message, language and idea, form and content, becomes particularly apparent when his style is examined. Ever fresh, poised and inventive, Gibbon exercises masterly control over his diction, allusions and quotations, his turns of phrase and figures of speech. Of course, the ever urbane Gibbon often takes delight in pure wit. Discussing the younger Gordian he writes:

> Twenty two acknowledged concubines and a library of sixty two thousand volumes, attested the variety of his inclinations; and from the productions which he left behind him, it appears that

the former as well as the latter were designed for use rather than for ostentation.

A footnote adds, 'By each of his concubines, the younger Gordian left three or four children. His literary productions were by no means contemptible.' There are plenty of mere conceits like this, but Gibbon's verbal touches more commonly hint at a serious point, as when we are told of Justinian's last years that 'his infirm mind was devoted to heavenly contemplation'. The implied equation of piety and senility says more than a page of analysis. Gibbon is particularly fond of using bathos and zeugma to strip pomposities and to deflate pretensions. Professors of literature, notably Harold Bond, have offered full analyses of Gibbon's use of language, his prosody, his three-piece sentence structure and the internal rhythms of his paragraphing; and there is no room to pursue such an approach here.

It must be stressed, however, that Gibbon's style comprises more than a mere felicity with phrases, a capacity to amuse or wound with a word: it is the vehicle of an ironic vision. Byron dubbed Gibbon the 'lord of irony', and rightly so. His ever-fastidious diction and balances, the antitheses and paradoxes of his clauses, all articulate a habit of irony which reaches to the heart of his philosophy of history, his interpretative bent. As Peter Gay says, Gibbon deployed irony because he saw the irony in the history of Rome.

Irony is basically saying one thing while meaning another; but such a formula ironically implies a simplicity which Gibbon's own more cunning usages commonly belie. For he frequently wields ironical turns of phrase not to establish the unsaid, but to unsettle the reader and hold him in suspense. Through fixing multiple layers of insinuation, innuendo and hidden meaning, Gibbon usually affirms neither one interpretation nor its simple opposite, but alerts the reader to a range of options; the final verdict rests in the reader's court. This technique is vital to Gibbon's subtle vision of history as a performance enacted before an audience, and we must never forget how passionately Gibbon loved the theatre, or the centrality of theatre symbolism for Augustan culture. The plot and script are composed of the evidence which has come down from the past. The historian has the power which is the prerogative of the director, but also the duty of fidelity to his script when setting the actors in motion. He may then shape and interpret, but it is finally the characters who perform and the audience who judges the performance.

Suspense is Gibbon's dramatic forte. He does not believe that the historian's business is to trade in solutions; history is not an answer-book. Rather it is the play of reality upon the world's stage. Gibbon constantly reminds us that history is a show; all we know is seen through other people's eyes, or filtered through someone's mind. Rumour, report and recollection fill the channels of our knowledge. No knowledge is unmediated, and in the *Decline and Fall* we are constantly kept aware of the reality of illusion itself, for Gibbon's protagonists are described as 'actors', wearing 'masks', playing their 'parts' – some better than others. Thus the artifice of the life of politics, the plots of power, form the great preoccupations of the historian of the Roman Empire, partly because Gibbon considered them the preoccupations of its emperors.

Gibbon's use of 'actor' is frequently pejorative. His characters are often portrayed as 'artful', masking their true intentions, weaving webs of illusion; rarely what they seem, they can *par excellence* smile and be villains. Sometimes Gibbon means to unmask a hypocrite. A jaundiced portrayal shows Augustus as a double-dealer, constantly exploiting his role, turning the Roman constitution into *trompe l'oeil*, and thereby eroding real virtue in the empire. Diocletian had evidently studied at the same drama-school, and turned state theatre into masquerade:

> Like the modesty affected by Augustus, the state maintained by Diocletian was a theatrical representation; but it must be confessed that, of the two comedies, the former was of a more liberal and manly character than the latter. It was the aim of the one to disguise, and the object of the other to display, the unbounded power which the emperors possessed over the Roman world.

Sometimes, however, Gibbon's point is that acting is indeed reality. Gibbon's dominant metaphor portrays the world as a theatre: all play parts, and it would be self-deception to believe that by stripping off one mask another will not be found beneath. The notion that the heroes of history are actors is not necessarily cynical or even pessimistic. Civilization may be a play, but it is a desirable one, so long as all play the game by the rules. Role-playing in public – enacting the noble prince, magnanimous victor, man of justice, courageous soldier – provides stability and predictability in a dangerous world where the lone individual is readily lost, the lost man becomes desperate, and the desperado turns dangerous. It is when conscious role-playing yields to self-deluding fantasy (as with the hallucinations of hermits in the early history of Christianity) that the play is abandoned and the alarm-bells ring.

In any case, it must, methodologically speaking, be the job of the historian, peering down into the deeps of time, to treat the past as a world of masks, public meanings and symbolic transactions. Lacking Momus' glass, the historian has no certain way of penetrating beneath the surface of actions and professions. Since the nineteenth century, Gibbon has been widely accused of presenting his *dramatis personae* as stock figures, types not individuals, mere poseurs, displaying ham gestures within a rhetorical syntax. They are constantly blushing with shame, dropping tears, bending the knee; theirs is the histrionic repertoire of the classical French theatre. Thus Gibbon is criticized for having a shallow psychology or none at all. Later historians, by contrast, especially those inspired by Romanticism, wrote as if they were privy to men's inner secrets, a tradition inherited this century by psycho-history. But Gibbon was diffident about plumbing motivation and more interested in charting the consequences of action.

It is Gibbon's way to present the reader with choices of interpretations, above all pairs of contrasting renderings of an action. Clovis's wife, he tells us, found it in 'her interest, as well as her duty', to achieve the conversion of a pagan husband, and Clovis insensibly listened to a voice of 'love and religion'. Gibbon here offers, very properly, a pluralist reading, as befits any historian aware of our ignorance about the *moeurs* of the Merovingians. Frequently his accounts of actions are supported by braces of antitheses: 'the pride or the policy', 'the piety or the avarice', 'the superstition or the greed', 'the courage or the fear'.

This presentation of psychological opposites is not just a stylistic twitch which we may applaud or deplore as we choose. Nor is it predominantly – as it had been for Gibbon's towering mentor, Tacitus – a vehicle for the controlled and concise expression of savage indignation against the corruption of the times or the depravity of men. Gibbon profoundly admired Tacitus, 'the first of historians who applied the science of philosophy to the study of facts'; and occasionally he has a truly Tacitean bite – 'Mascezel was received at the court of Milan with loud applause, affected gratitude, and secret jealousy' – or a Tacitean contempt ('the tender respect of Augustus for a free constitution which he had destroyed'). But Gibbon is not Tacitean in temper.

Peter Gay, who has written with wonderful acuity on Gibbon's irony and his debt to Tacitus, suggests that when we encounter Gibbon's habitual coupling of 'higher and lower motives' we are left with no alternative. We are directed into the lower, 'unpleasant, generally cynical', option: the avarice, not the piety, as it were. Yet Gibbon, who sports a lofty detachment alien to Tacitus, speaks with a voice more sceptical than cynical, and may be offering the reader something more puzzling, more suggestive. He toys with these ambiguities not to make a point, but to unmake it, to open up debate, to indicate the radical contingency of performance both on the historical stage and in the processes of historical interpretation. Gibbon aims to provoke doubt, not loathing; unlike Tacitus, he is a rabid partisan only for the party of humanity. His strategy continually forces the reader to examine his own stock responses and prejudices, and confront his rationalizations. His prose probes into the play of mind upon mind, and he is always fascinated by the power of ideology, the rationalizations which language affords; in a playful letter he states that 'the principal use, I know in human reason is, when called upon, to furnish arguments for what we have an inclination to do' – a thorough-going Humean rendering of reason as passion's slave. In other words, irony is the medium whereby the philosophical, even the sceptical, historian expresses the radical uncertainty of things and holds the reader in suspense even while taking him into his confidence. Gibbonian irony incorporates the salutary lessons of the Pyrrhonists while saving the very project of making the past make sense.

History is never simply as it seems. More goes on than meets the eye, but often the play, the bias, the modulation of reality is subtle, hard to pin down, and more properly hinted at than stated. Take the very first page of the *Decline and Fall* whose first lines were quoted earlier when I read off its surface meaning.

In the second century of the Christian aera, the empire of Rome comprehended the fairest part of the earth, and the most civilized portion of mankind. The frontiers of that extensive monarchy were guarded by ancient renown and disciplined valour. The gentle, but powerful, influence of laws and manners had gradually cemented the union of the provinces. Their peaceful inhabitants enjoyed or abused the advantages of wealth and luxury. The image of a free constitution was preserved with decent reverence. The Roman senate appeared to possess the sovereign authority, and devolved on the emperors all the executive powers of government.

Investigated closely, the passage dissolves into a sea of doubt; no ultimate meaning establishes itself. Gibbon is calling our attention to the fact that reality has been supplanted – or maybe supplemented? – by appearances ('the image of a free constitution': where does that leave real freedom?). He is darting paradoxes at us (Rome is a 'monarchy' and yet the Senate still 'appeared to possess the sovereign authority'), and offering in condensed form alternative readings (the Romans 'enjoyed or abused' the advantages of wealth and luxury: which was it? is there *really* any difference?). Here as elsewhere Gibbon forces us to weigh the meaning of acts, by drawing attention to such words as 'studied', 'affected', 'seemed' and 'professed', with all their inherent ambiguities. Not least, what do some of the terms mean? 'Luxury' – just how pejorative is this then-emotive word in Gibbon's vocabulary?

The Gibbonian past seems lucid, but is ultimately opaque. We see various layers and, lifting the veil, we perceive yet more. Which are more 'real' than others? It may be the appearances and masks which make civilization possible; for Gibbon, like Hume, believed that mankind was governed by opinion. Not least, Gibbon is wielding the charged language of irony as a means of reminding us of the unfathomable complexity of motivations, explicit and 'insensible'. It would be fruitless to try to clinch, as in a Newtonian experiment, the nature of Constantine's motives (was his conversion sincere or a politique act of state?). So Gibbon weaves his own interpretation into the very texture of his words, tone and narrative. Sacred history had seen the finger of Providence pointing the plot of history. Nineteenth-century philosophers professed to see reason working *in* history. But Gibbon personifies reason standing *outside* history, sustained intelligence in a great head of learning, controlling, judging and always interpreting.

As I noted in the Introduction, critics have complained that Gibbon's history lacks pure analysis, formal assessments of motives, systematic evaluations of primary causes and secondary laws. But this critique is surely misguided, the result of a search for particular passages in the *Decline and Fall* where pure interpretation is woven into the design of Gibbon's tale. We may apply to Gibbon himself the tribute he bestows upon Tacitus' 'immortal work', that 'every sentence ... is pregnant with the deepest observations and the most lively images.'

4 Power

The *Decline and Fall* is at bottom a study of the workings of power. The unmoved mover of history is not justice or morality, nor a law of progress, nor even the index finger of God. There is no transcendental meaning; there was no state of nature, there is no ultimate goal to which all things are tending, there will be no withering away of the state. Gibbon had no truck with metaphysics. History is just the outworking of energy and conflict among peoples organized into political units. Gibbon is a realist – even, one might say, a Machiavellian: 'wars, and the administration of public affairs are the principal subjects of history.' Through perusing his book, Gibbon tells his readers, 'the most voracious appetite for war will be abundantly satisfied.'

Gibbon was not power-mad; he merely faced its reality. For a long while – for perhaps a thousand years – Rome was king fish in the Mediterranean. She destroyed Carthage; she conquered the natives of Italy, Spain, Gaul and Dalmatia; she vanquished the civilized states of Greece and Egypt; she warred with honour against the might of Persia. Ruling, as Virgil stated, was the art of the Romans: they did it very well. Yet Gibbon's story also shows the decay of Roman arms and the 'deluge of barbarians', until Byzantium finally fell to the siege-engines of the Turk.

Wherein then lay power? Immediately, of course, in military prowess, the organization of the army, superiority in technology and tactics, and most importantly in military discipline and morale. Historians had to know their weapons and tactics: 'the captain of the Hampshire grenadiers,' Gibbon teased, 'has not been useless to the historian of the Roman Empire'; and he delighted in imagining the relative military capabilities of the Roman legion, with its disciplined infantry and its sword and buckler tactics, matched against the mad naked bravery of the Goths or the fleet cavalry of the Persians and the Saracens.

But Gibbon knew that power did not gleam from swords and soldiers alone; they were merely its expressions in the long run

94

and the last resort. It was, after all, 'ancient renown', Gibbon stressed, or in other words *ideology*, which defended the boundaries in the second century. Power lay in the texture of political societies, in the amazingly intricate interplay between the populace, their leaders and military capacity. And, following the classical historians and their later interpreters such as Harrington and Walter Moyle (discussed earlier in Chapter 1), Gibbon advanced a coherent thesis to explain what had made Rome mistress of the Mediterranean littoral by the first century after Christ, what had given Rome that mission of ruling, that temper of government compounded of justice, clemency, toleration, civic virtue, patriotism and cosmopolitanism, which he so warmly admired.

It was a theory which asserted the connexions between the republican mode of government and successful military expansionism. Rome, originally a monarchy, had chafed under its despotism. Once the kings were overthrown, as first Livy and later Machiavelli had emphasized, the people enjoyed liberty, and citizens saw themselves as holding a stake in their own city. Property-holders participated in the political process, some more effectively than others. Because office-holding remained essentially elective and annual, power did not become the hereditary right of any individual or hereditary clique. All fought as citizen-soldiers in Rome's wars. The inevitable internal social and class tensions of the expanding state were successfully checked by a mixed constitution, and by an aggressive, inclusive chauvinism; military success in turn ensured general prosperity. Patriotism, justice and liberty flourished during the glorious days of the swelling republic, creating a momentum which served the young empire. Learning from Montesquieu, Gibbon judged ardent public involvement the key to Rome's success:

> That public virtue, which among the ancients was denominated patriotism, is derived from a strong sense of our own interest in the preservation and prosperity of the free government of which we are members. Such a sentiment ... rendered the legions of the republic almost invincible.

Yet it was a delicate equilibrium. Small was effective. But the small state expanded, and, the grander Rome grew, the less easily maintained was the civic, political freedom upon which success had been built. Rome prudently, generously even, assimilated its conquered tribes; but beyond Italy such people did not initially achieve citizen status, nor could they share the original political fervour. Broadening frontiers and continual war led to armies manned by provincials, and Rome's ability to incorporate its clients and conquered peoples was indeed part of its genius:

> The grandsons of the Gauls who had besieged Julius Caesar in Alesia commanded legions, governed provinces, and were admitted into the senate at Rome. Their ambition, instead of disturbing the tranquillity of the state, was intimately connected with its safety and greatness.

Yet this also generated deep problems, especially among the common soldiery. They could not be motivated by the old civic patriotism, but had to be spurred by religion, honour and military discipline. There emerged a breed of professional soldiers which inevitably, starting with Marius and Sulla, and culminating with Pompey in Bithynia, Crassus in Parthia, and Caesar in Gaul, spawned adventurer-generals chancing their arm with military might against the political authority of the Senate. The danger of the army dictating to the Senate, and not vice versa, grew. Rifts began to appear between civil authority in Rome and the real, military might in far-flung provinces. Civil war resulted; Caesar was assassinated, and two generations of turmoil were finally closed by Augustus' inception of empire.

As discussed above in the first chapter, Gibbon, like so many of his fellow Englishmen, saw reflected in the politics of the Roman Republic the qualities ideally displayed in Georgian England and the British constitution. Liberty, that precious but precarious jewel, was above all protected by law and custom, checks and balances. In both polities 'a rational freedom' – so easily sundered by the absolute power of the prince or plebeian mob – was safeguarded by a 'mixed constitution', the delicate interplay of political interest groups, the astute division of powers, the contest between the various limbs of the body politic:

> a martial nobility and stubborn commons possessed of arms, tenacious of property and collected into constitutional assemblies form the only balance capable of preserving a free constitution against the enterprises of an aspiring prince.

Gibbon believed that the Roman republic had possessed the right alchemy: 'The ... struggles of the patricians and plebeians had finally established the firm and equal balance of the constitution, which united the freedom of popular assemblies with the authority and wisdom of the senate.'

This theory of the virtue of a balanced constitution was not applicable only to republics in the strict sense; Gibbon and his contemporaries thought it perfectly compatible with constitutional monarchy. The tragedy of Rome was that it never grew into an authentic constitutional monarchy; the evil imperializing genius of Augustus placed this delicate balance in jeopardy.

Gibbon characterizes Augustus as a 'crafty tyrant'. By that he means that largely for personal advantage Augustus sapped the virtue of the state, and began the extinction of that robust political life which had sustained the republic. Temperamentally a 'tyrant by stealth', Augustus chose to 'reign under the venerable names of ancient magistracy, and artfully to collect in his own person the scattered rays of civil jurisdiction'. 'Sensible that mankind is governed by names', he cunningly preserved all the forms of the old republican constitution, reinstated the Senate with its powers, and retained such time-hallowed offices as the consul's. But in effect he reduced them to ceremonial charade. The Senate became his rubber stamp, 'a tractable and useful instrument of dominion'. Through Augustus' political vision, Rome was to be reduced to 'an absolute monarchy disguised by the forms of the commonwealth':

> The masters of the Roman world surrounded their throne with darkness, concealed their irresistible strength, and humbly professed themselves the accountable ministers of the senate, whose supreme decrees they dictated and obeyed.

Fatigued by decades of warfare and civil strife, the political classes relaxed and resigned their political future. Freedom leaked away. It was only a matter of time until Rome itself fell into a decline. At first the process was invisible. Indeed, as Gibbon's opening pages so clearly exemplify, under the benign reigns of the five good emperors of the second century, Rome seemed to be enjoying its zenith, content within its well-established frontiers. The *pax romana* was secured:

> The principal conquests of the Romans were achieved under the republic, and the emperors for the most part, were satisfied with

preserving those dominions which had been acquired by the policy of the senate, the active emulation of the consuls, and the martial enthusiasm of the people.

Thus, as Gibbon explained, the peaceful empire presented a fair prospect:

> The image of a free constitution was preserved with decent reverence. The Roman senate appeared to possess the sovereign authority, and devolved on the emperors all the executive powers of government. During a happy period of more than fourscore years, the public administration was conducted by the virtue and abilities of Nerva, Trajan, Hadrian and the two Antonines.

But if it was Rome's summer, it proved an Indian summer. Though for 220 years 'the dangers inherent to a military government were, in a great measure, suspended,' nevertheless imperial peace proved insidious:

> It was scarcely possible that the eyes of contemporaries should discover in the public felicity the latent causes of decay and corruption. This long peace and the uniform government of the Romans, introduced a slow and secret poison into the vitals of the empire. The minds of men were gradually reduced to the same level, the fire of genius was extinguished, and even the military spirit evaporated.

And the process of corruption went on. The imperial monarchy encouraged apathy, weakening the resolve and the nerve of the old political orders. While absolutism remained 'enlightened', the effect was not disastrous. Yet 'the happiness of an hundred millions depended on the personal merit of one or two men, perhaps children', and not all despots were benevolent; after all, even the first century had witnessed in rapid succession 'the dark, unrelenting Tiberius, the furious Caligula, the stupid Claudius, the profligate and cruel Nero, the beastly Vitellius, and the inhuman Domitian'. Gibbon profoundly distrusted empire.

Gibbon's repudiation of absolutist government has often been obscured. Fuglum has written, for example, of Gibbon's 'cult of the Roman Empire' ('a longing for a state of things that never was and never can be, a nostalgia for an imagined golden age'); and Momigliano has argued, confusingly I believe, that Gibbon 'shared

Voltaire's indecision as to whether constitutional government or the enlightened despot were the better regime for a state'. More directly, Leslie Stephen maintained that for Gibbon, a shallow rationalist, the 'ideal state of society is the deathlike trance of an enlightened despotism':

> He does not sympathize with the periods marked by vehement ebullitions of human passion breaking down the frozen crust of society, evolving new forms of religion, art, and philosophy, and in the process, producing struggles, excitement, and disorder, but with the periods of calm stagnation, when nobody believes strongly, feels warmly, or acts energetically. A peaceful acquiescence in the established order, not an heroic struggle towards a fuller satisfaction of all human instincts, is his ideal. Equilibrium, at whatever sacrifice obtained, is the one political good; and his millennium can be reached rather by men ceasing to labour than by their obtaining a full fruition. In all which, of course, Gibbon is the representative man of his time and class.

Stephen's interpretation not only travesties both Gibbon and his times, it more or less reverses Gibbon's political ideology and his historical understanding of power. Stephen tells us, for instance, that Gibbon lauds equilibrium, but the historian famously states: 'all that is human must retrograde if it does not advance.' In fact, the key to Gibbon's political vision was not 'calm stagnation' but 'freedom': 'Freedom is the first wish of our heart; freedom is the first blessing of our nature.'

Freedom, in Gibbon's eyes, was 'the happy parent of taste and science', it was 'the source of every generous and rational sentiment', the 'most powerful spring of the efforts and improvements of mankind'. Liberty is the Ariadne's thread to Gibbon's views as a historian and political observer. No tendentious theories of Oedipal struggle need be invoked to recognize that he personally fretted under paternal authority. Looking back in his *Memoirs*, he depicted his grandfather in Putney acting as the 'oracle and tyrant of a petty kingdom' (some allowance must be made for young Gibbon's fevered imagination!); he talked of the tyranny of the 'cruel and capricious paedagogue' during the days when he suffered 'the dependence of a schoolboy'; and saw childhood, marked by powerlessness, as a state of 'servitude' – and 'I never could understand the happiness of servitude.' The only words of his mother he could remember were uttered on the occasion she packed him off to school at the age of nine. He must, she told him, 'learn to think and act for myself': it was the best piece of advice he ever had.

He hated being reduced to boyish dependence once more when exiled to Lausanne by that arbitrary father who, as Jordan puts it, then:

> called him home to break the entail, he forbade his marriage, he remarried without telling him, he pushed him to publish the *Essai sur l'étude de la littérature*, he enlisted him in the Hampshire militia, he insisted Gibbon live at Buriton, he reneged on his commitment to finance the grand tour, he tried to force his son into Parliament [etc.].

Gibbon, as we have seen, resented the years spent during his twenties still under his father's tutelage, when he felt a growing yearning to be 'master in my own house'. Jordan says Gibbon 'hated his father'; this is to exaggerate, but his relations with his father certainly awakened a profound loathing of dependence and arbitrary power, which perhaps in part explains his attack on Warburton in 1770, when he melodramatically accused the prelate of exercising intellectual 'despotism' as the 'Dictator and tyrant of the world of literature', doling out 'infallible decrees': 'In a land of liberty, such despotism must provoke a general opposition. I too ... was ambitious of breaking a lance against the Giant's shield.'

All of Gibbon's experiences during early manhood led him to champion political liberty. After a childhood steeped in family Toryism, he turned himself, with the help of reading Locke, into a good Whig. Returning from his Grand Tour, he sought the opportunity to visit Holland ('a country, the monument of freedom and industry'). His fragmentary essay on Charles VIII of France, written in 1761, argued, following good Whig rhetoric, that the 'only title [to a throne] not liable to objection is the consenting voice of a free people'. And his history of the Swiss was designed to be the story of a 'brave and free people':

> The Swiss cherishes his family and his companions, respects religion and the laws, scorns fatigue, braves death, and fears only infamy. Liberty is dear to him, and this independence, which breeds equality of fortune and sense of his strength, is the first spring of his soul.

Perturbed by the lack of autonomy suffered by the burghers of Lausanne, ruled at a distance by the Bernese oligarchy, he penned his anonymous and unpublished 'Lettre d'un Suédois sur le gouvernement de Berne'. The *pays de Vaud* was in many ways an admirable region, Gibbon admitted. 'What do you lack? Liberty: and deprived of liberty, you lack everything' ('prosperity,' he priggishly informed his hosts, 'is the counter argument of a slave').

Service in the militia gave Gibbon experience of that active compound of military and political involvement he admired in ancient Rome ('what I value most, is the knowledge it has given me of mankind in general, and of my own country in particular'). An MP from 1774, he experienced at first hand the old Roman senatorial thrill of having not just a stake but a direct vote in the government of his country, and he learned, he believed, the crucial wisdom of politics ('the eight sessions that I sat in Parliament were a school of civil prudence, the first and most essential virtue of an historian'). True, Gibbon's own political career had its ignominious side. He never once opened his mouth in all his years in the Commons, and eventually he became a placeman as a Commissioner of the Board of Trade under Lord North. It would, however, be somewhat unfair to call it a sinecure; Gibbon attended 117 meetings in three years. He was not quite the political toady that his detractors sometimes portray, for he seems, like a good backbencher, to have kept his political conscience ('he votes variously as his opinion leads him,' noted Horace Walpole, who found such behaviour whimsical). Casting himself into the senatorial mould, Gibbon idolized the heroes of the republic, the champions of Roman freedom such as Cicero ('I tasted the beauties of language, I breathed the spirit of freedom, and I imbibed from his precepts and examples the public and private sense of a man'). In short, Gibbon claimed, 'Although I have devoted myself to write the annals of a declining monarchy, I shall embrace the occasion to breathe the pure and invigorating air of the republic.'

For Gibbon the tragedy of empire was that it sapped freedom. That necessarily brought a trail of utter disaster: 'the succession of five centuries inflicted the various evils of military license, capricious despotism and elaborate oppression.' Precisely how and why was imperial absolutism so disastrous? Power vested in the person was intrinsically inimical to freedom: 'unless public liberty is protected by intrepid and vigilant guardians, the authority of

so formidable a magistrate will soon degenerate into despotism.'
For one thing, it removed any curbs upon power. The Emperor,
campaigning through the vast terrains of the empire, came to think
that his essential power-base was the loyalty of his own troops;
his relations with the Senate in Rome readily reduced themselves
to questions of the Emperor's control over the fierce praetorian
guards, stationed around Rome, who in effect operated as his personal bodyguard. Thus emperors began to command military power
in their own name, and constitutional authority – itself increasingly
a cipher – became fatally severed from military might. Soldiers no
longer looked to the Senate and people of Rome for orders and
allegiance. Generals and emperors were their paymasters, their
hopes for the future, and they felt themselves to have little stake
in the city:

> In the purer ages of the commonwealth, the use of arms was
> reserved for those ranks of citizens who had a country to love,
> a property to defend, and some share in enacting those laws,
> which it was their interest as well as duty, to maintain. But in
> proportion as the public freedom was lost in extent of conquest,
> war was gradually improved into an art, and degraded into a
> trade.

Reduced thus to mercenaries, legionaries looked to their leaders
for reward and advancement; similarly the provincials lost their
involvement with Rome itself:

> Their personal valour remained, but they no longer possessed
> that public courage which is nourished by the love of indepen-
> dence, the sense of national honour, the presence of danger, and
> the habit of command. They received laws and governors from
> the will of their sovereign, and trusted for their defence to a
> mercenary army.

The progress of the empire thus became a chronicle of generals
eagerly bidding for power. The Romans themselves opted out of
fighting, and the generals ceased to be Romans, or even Italians;
hence they lacked loyalty to the old political machinery back in
the capital and its 'ancient renown'. Generals made themselves masters of the army; nor was it long before the army in turn mastered
its master, and created 'the imperial slave'.

The old senatorial classes supinely permitted this transformation to occur without a struggle. They abdicated their responsibilities: 'as long as they were indulged in the enjoyment of their baths, their theatres, and their villas, they cheerfully resigned the more dangerous care of empire.' For Gibbon the demise of these *optimates* delivered an object lesson to his own times, evincing the importance of vigilance against corruption. Senators had taken the easy way out. By allowing the emperors a free hand, patricians had been permitted to indulge in the masquerade of power, without any requirement to exercise or sacrifice themselves for the public good. They were rather like Gibbon's idle Oxford tutor, Dr Winchester, who 'well remembered that he had a salary to receive, and only forgot that he had a duty to perform'. The governing classes of Rome, Gibbon believed, acquiesced in their reduction to parasitism and political impotence. They paid for their apathy at the close of the third century.

Montesquieu's political theory, upon which Gibbon drew, predicted that monarchy readily declined into despotism. This is what happened to Rome. Augustus had preserved the veneer of the old constitution, the mask of freedom and legality. A plateau of peace was maintained by the five benign emperors – yet even they, Gibbon mused, might be troubled by the 'melancholy reflection' that the time might come when 'some licentious youth, or some jealous tyrant, would abuse, to the destruction, that absolute power which they had exerted for the benefit of the people'. That peace was shattered by the military anarchy of the mid-third century, which provoked Diocletian to abolish even the husks of the constitution.

In Gibbon's interpretation, Diocletian saved the empire only by militarizing and bureaucratizing it. He introduced conscription, and massively reinforced the army, while curbing the powers of the generals. Furthermore, he began to transform the empire into an oriental despotism. The trappings of a court were developed to increase the aura of the Emperor. He launched a cult of the Emperor's own person. The imperial bureaucracy was cultivated to protect, distance and magnify the ruler. All such governmental reforms in turn required greater expense, and thus new fiscal exactions and higher taxation, which fell principally upon the cultivators of the land; by consequence the wealth-base of the empire came to be undermined. As a result, more were reduced to servility ('the natural propensity of a despot [is] to sink all his subjects to the same common level of absolute dependence'), many flocking to Rome itself, where the free corn doles were attractive. Gibbon does not pursue the purely economic causes of Roman decay – he has little to say about the so-called crisis of a slave-owning economy, of which later Marxist historians have made so much. His stress is primarily political. Higher land and poll taxes were evils because they extended dependency upon imperial power – an analysis commonly applied also to Georgian England.

With Diocletian and his successor Constantine, Gibbon argued, the principate gave way to Caesarism. And a new imperial style was increasingly adopted. Elaborate and ostentatious dress, hairstyles, cosmetics, ceremonial and royal insignia reinforced the majesty of power and abasement of the subject. Thus tyranny entered with 'an air of softness and effeminacy'. Gibbon notes with distaste how Diocletian was:

> represented with false hair of various colours, laboriously arranged by the skilful artists of the times; a diadem of a new and more expensive fashion; a profusion of gems and pearls, of collars and bracelets, and a variegated flowing robe of silk, most curiously embroidered with flowers of gold.

Under Diocletian and Constantine Roman courage gave way to Eastern decadence:

> The manly pride of the Romans, content with substantial power, had left to the vanity of the East the forms and ceremonies of ostentatious greatness. But when they lost even the semblance of those virtues which were derived from their ancient freedom, the simplicity of Roman manners was insensibly corrupted by the stately affectation of the courts of Asia.

Oriental despotism, with its distant echoes of Alexander, rang the changes in the theatre of power. 'The philosophic observer', observed Gibbon, might now have mistaken the system of Roman government for a 'splendid theatre', filled with 'players of every character and degree, who repeated the language and imitated the passions of their original model'. Sometimes, indeed, the new atmosphere of the court appeared more like a brothel than a stage.

For Gibbon, Constantine's orientalization of imperial power was an unmitigated disaster. With 'a new capital, a new policy, and a new religion', there could be no check on the adoption of Eastern forms of rule. 'The sovereign advised with his ministers, instead of consulting the great council of the nation.' The history of the Eastern empire, as Gibbon judged it, was one long slide into servility, superstition and sexual abuse. Government became theocratic, and power fell into the hands of women, bishops, slaves and functionaries. This new Byzantine polity handed power for the first time in Roman history to a class of eunuch-ministers, owing their position solely to the grace and favour of the Emperor himself – it was, in Gibbon's eyes, a peculiar monstrosity. Inevitably emperors' wives completed the emasculation of Roman rule, and precipitated a descent into sexual politics. Gibbon's chauvinist panorama of the sordid centuries of Byzantine politicking ('the indolent luxury of Asia') is filled with backstairs and bedroom politics, petticoat government and the trading of sexual favours by imperial wives and princesses, leading ultimately to the murder of emperors by their oversexed consorts, destroyers of men. Byzantine rule meant the seduction of the empire by eunuchs and whorish queens, such as the notorious Empress Theodora or Antonina, the wife of Belisarius. Theodora's mother was a 'theatrical prostitute', and she in turn cuckolded her husband; Antonina vied with Theodora in the 'jealousy of vice', before her final reconciliation in 'the partnership of guilt'. Or take Theophano, the wife of Romanus II, son of Constantine VII. First suspected of having murdered her father-in-law, this 'woman of base origin, masculine spirit, and flagitious manners' finally slew her husband.

The reign of Constantine had driven yet another nail into the coffin of political liberty and patriotism: for it signalled the politicization of Christianity. In the next chapter I shall analyse Gibbon's interpretation of the growing significance of religion for the destiny of Rome. Here it is worth examining his view of the use and abuse of religion within the commonwealth.

Whatever Gibbon's own religious sensibilities, it would be a mistake to represent him as implacably hostile to religion as such. He delivers a sympathetic account of the pagan polytheism typical of the Roman republic and the early years of the empire. There were many gods, and, perhaps like football teams nowadays, the diversity of deities provided the varied peoples of the empire with focuses of loyalty. At the same time the official Roman gods and cults rendered the fortunes of the state the object of people's hopes, and provided sanctions for civil duty. Thus civic religion was a source of social strength. Universal tolerance, enforced by the magistrates, prevented religious discord.

That idyll ended of course with the rise of Christianity. Mono-theistic, zealous and intolerant of other gods, Christianity threatened to found a state within the state, and thus generated civil strife. As Gibbon saw it, however, Christianity might have remained one of many religions within the empire, and never divided it. The cross-ing of the Rubicon came with Constantine's conversion. Historians had traditionally disagreed about how to interpret this act: was it (as of course ecclesiastical historians treated it) a signal victory for Providence? Or the act of a grossly superstitious man, clutching at any promised straw of divine favour? Or rather the deed of a *politique*, a crafty manipulator, who saw that imperial absolutism could best be served by the installation of an absolutist creed? One God, one emperor, one faith – what better reinforcement of hier-archical, indeed theocratic, authority?

Gibbon inclined towards the last view. He certainly insisted that Christianity's convictions proved wonderfully convenient to the development of the theocratic monarchy of the Eastern empire. The more the Emperor was loaded with the trappings of divinity, the easier to view the political structure as a simple chain of obedience leading directly up to the Emperor and thence to God Almighty. Once 'the clergy preached the doctrines of patience and pusillanimity', service not liberty became the political watchword. Christianity, with its ethic of 'passive obedience', magnified all the hieratic, servile tendencies of oriental despotism.

It is often said that Gibbon's extreme distaste for the Byzantine state blinded him to its real strength and success. He certainly paints it as a sinkhole of stagnation, corruption and sycophancy. Yet as Runciman, Vryonis and other commentators have pointed out, the empire centred at Constantinople survived for well over a thousand years. During periods of expansion, in particular under the general-ship of Belisarius, it temporarily regained many of the lost North African territories, and recaptured parts of Italy for good. Not least, Constantinople remained the largest and busiest city in the Mediter-ranean world. Is Gibbon's vision of abject, effeminate weakness merely a phantom of prejudice?

Perhaps critics of Gibbon's analysis of Byzantinization of power do not fully recognize his real objection to the Eastern empire. His aim was to show that Byzantinism of power actually strengthened

the position of the Emperor, but at the expense of the political
health of the commonwealth itself. As a political analyst, Gibbon
– unlike Machiavelli in *Il Principe* – never narrowed his field of
vision to the aggrandizement of the individual ruler. Nor did he
accept the casuistry which identified the vitality of the state with
the good of the prince, for he believed that the wealth of princes
was generally bought at the expense of their people. In Savoy, he
saw a village dying of cold and hunger in every gilded moulding
of the palace at Turin. For Gibbon, there was no merit in the survival
of dynasties unless they conferred upon the people the dignity of
warlike courage, civic pride and a flourishing artistic and intellectual
culture. All these were jeopardized by Eastern absolutism. Even
the happy years of the Antonines had borne testimony to this fact.
The great efflorescence of arts and philosophy of the late republic
came to a halt. Instead, there was cultural death: 'the minds of
men were reduced to the same level, the fire of genius was
extinguished.' The Byzantine period was far worse. A thousand
years passed without the emergence of a single poet, orator, historian
or moralist who met with Gibbon's approval. Byzantine history-
writing descended into puerile panegyric, and its intellectuals wasted
their breath upon sterile theological wrangling; the traditional
penchant of the Greek mind for vacuous metaphysics was now
harnessed to the service of Christian fanaticism.

Thus the rise of Caesarism destroyed Rome while, in the short
term, benefiting its emperors. Gibbon did not see this loss of political
virtue as inevitable; it was due to a specific failure of civil prudence,
which could have been avoided. Even under the empire, wise provi-
sion had been made, albeit temporarily and on a local basis, for
the kind of political participation which Gibbon thought preserved
freedom. He described Honorius' summoning of the assembly of
Arles, involving the praetors, the bishops and the major landholders:

> They were empowered to interpret and communicate the laws
> of their sovereign; to expose the grievances and wishes of their
> constituents; to moderate the excessive or unequal weight of
> taxes; and to deliberate on every subject of local and national
> importance that could tend to the restoration of the peace and
> prosperity of the seven provinces. If such an institution, which
> gave the people an interest in their own government, had been
> universally established by Trajan or the Antonines, the seeds of

public wisdom and virtue might have been cherished and propagated in the empire of Rome. The privileges of the subject would have secured the throne of the monarch; the abuses of an arbitrary administration might have been prevented, in some degree, or corrected, by the interposition of these representative assemblies; and the country would have been defended against a foreign enemy by the arms of natives and freemen. Under the mild and generous influence of liberty, the Roman Empire might have remained invincible and immortal; or, if its excessive magnitude and the instability of human affairs had opposed such perpetual continuance, its vital and constituent members might have separately preserved their vigour and independence.

Rome could have been saved by Parliament.

In view of this vision of the collapse of the public good under the later empire, it is not surprising that Gibbon had something favourable to say of the barbarians. As I shall explain more fully in Chapter 6, Gibbon's assessment of the hordes descending upon the Roman Empire in the fourth century and after was unromantic. He did not claim that primitive genius was manifest in their social habits or institutions; neither did he sentimentalize them as noble savages. Following his mentor and chief source Tacitus, however, he emphasized the freedom which their very backwardness engendered:

> A warlike nation like the Germans, without either cities, letters, arts, or money, found some compensation for this savage state in the enjoyment of liberty. Their poverty secured their freedom, since our desires and our possessions are the strongest fetters of despotism.

There were of course exceptions, but the exceptions proved the rule:

> 'Amongst the Suiones (says Tacitus) riches are held in honour. They are therefore subject to an absolute monarch, who, instead of intrusting his people with the free use of arms, as is practised in the rest of Germany, commits them to the safe custody not of a citizen, or even of a freedman, but of a slave. The neighbours

of the Suiones, the Sitones, are sunk even below servitude; they obey a woman.' In the mention of these exceptions, the great historian sufficiently acknowledges the general theory of government.

The political freedom of savagery, however, was not particularly to be prized, for it left the barbarians enslaved to every gust of passion and every turn of fortune. Unlike many contemporary political commentators such as Montesquieu and Boulainvilliers, Gibbon saw it as no part of his historico-political testament to explain how modern liberty had sprouted from the Teutonic forests and the feudal relations which had grown up there. Eventually, of course, the new barbarian nations had given birth to civilization and taste. But Gibbon laid his stress upon the appalling tardiness of the process: it had taken no less than a thousand years.

Gibbon was pre-eminently a political realist. In the real world, the alternatives were to dominate or be dominated. That was even a kind of personal truth; as he put it in a rather chilling observation which hints at Hegel's master–slave dialectic, 'such is the law of our imperfect nature, that we must either command or obey, that our personal liberty is supported by the obsequiousness of our own dependents.' Even if glory were empty it was better to be powerful than powerless, victorious than vanquished. The great could exercise mercy and teach the lessons of civilization. But naked power was an evil. When he compared Frederick the Great or Catherine the Great to the emperors of Antiquity ('a Julian, or Semiramis, may reign in the North'), he meant to flatter neither. It is sometimes asked why then this enemy of absolutism became so overwhelmingly hostile to the French Revolution. I shall touch upon Gibbon's understanding of that event below; but it can here be said that he consistently viewed the prospect of mob rule and demagogy with no less horror than the naked power of the prince. The 'licentious freedom of the many' could be only 'an evil'. Like Burke ('I admire his eloquence – I approve his politics, I adore his chivalry') he feared extremism and knew that absolute power, whether popular or princely, corrupted absolutely: under 'the wild theories of equal and boundless freedom', he expected the dissolution of the social contrivances, the checks and balances, which ensured the moderating influence of civilization:

This total subversion of all rank, order, and government, could be productive only of a popular monster, which after devouring everything else, must finally devour itself.

These words reiterate one of his fundamental complaints against the oriental absolutism of the Eastern empire: its equalization of all before the ruler at the level of slavish abjection.

5 <u>Religion</u>

When Hannah More, the evangelical bluestocking, heard of Gibbon's death she exclaimed: 'How many souls have his writings polluted! Lord preserve others from their contagion!' She was of course recoiling − like all the crusading critics who took up their pens − against Gibbon's habitual cynicism towards Christianity. It is an antagonism constantly bubbling up through the sneers, *doubles entendres* and irony. 'Of the three popes,' Gibbon notes, amazed at the crop of pontiffs and antipopes parading at the Council of Constance during the Great Schism, 'John the twenty third was the first victim: the most scandalous charges were suppressed; the vicar of Christ was only accused of piracy, murder, rape, sodomy, and incest.' The papal Johns were indeed a rum lot. Of John xii, Gibbon records, eyebrows raised, 'we read with some surprise ... that his rapes of virgins and widows had deterred the female pilgrims from visiting the tomb of St Peter, lest in the devout act, they should be violated by his successor.' At least he found these sybaritic and salacious popes rather more human than monks and mortifiers of the flesh. And towering over them all was Simeon Stylites, who dwelt on top of a pillar for thirty years: 'the name and genius of Simeon Stylites have been immortalized by the singular invention of an aerial pennance.' But perhaps even he was out-Simeoned by another pillar of righteousness, 'the learned Origen', who, pricked by sexual temptation, 'judged it the most prudent to disarm the tempter', and castrated himself. Sometimes Gibbon's humour is marked less by malice or outrage than by bemusement at human folly. Martin of Tours set out to destroy by force all the idols and idolatry in his diocese. Poor Martin! 'The saint once mistook (as Don Quixote might have done) an harmless funeral for an idolatrous procession, and imprudently committed a miracle.' Dear reader, Gibbon is saying, who would believe all that these Christians have got up to? − or indeed, what they *believe*, as when he comments upon yet another pope, that he:

asserted, most probably with truth, that a linen which had been sanctified in the neighbourhood of [St Paul's body] or the filings of his chain, which it was sometimes easy and sometimes impossible to obtain, possessed an equal degree of miraculous virtue.

All good clean fun, and pious eighteenth-century Protestants hardly had a leg to stand on when they accused Gibbon of impiety, for they had been making the same kind of jibes against Catholic credulity and imposture for centuries.

Religion has traditionally been judged Gibbon's blind spot. He had not the slightest spiritual sympathies, Leslie Stephen contended, and many later historians have also portrayed him as a once-born worldling who lacked the finer strivings of the soul. And what he could not understand, he mocked. Certainly Gibbon's normal affability deserts him when he is depicting Christians; and Porson was not far wrong in treating Gibbon as a man of astonishing sympathies except when damsels were being ravished or Christians martyred. Hence it seems better to regard his treatment of Christianity as revealing less amused indifference than an intense if tortuous preoccupation with religion and its perils for individuals and society alike.

Gibbon battled with a religious demon from childhood. His infancy was darkened by the long shadow of the devout High Churchman, William Law, whose quest for personal holiness, to be achieved through austerity, offices, prayer, fasting and preferably chastity, had stained English practical divinity with Catholic mysticism and devotion yet again. His last compositions, warns Gibbon, are 'darkly tinctured with the incomprehensible visions of Jacob Behmen', and he affects surprise that 'Fanatics' such as Law 'who most vehemently inculcate the love of God should be those who despoil him of every amiable attribute'. In some ways Gibbon did, however, appreciate Law, despite his being 'clouded with enthusiasm', for he was a kind of pocket Pascal, severest on the hypocrisies and backslidings of fellow believers, exposing 'with equal severity and truth, the strange contradiction between the faith and practice of the Christian world'.

Gibbon never met Law, but he knew only too well his own formidable maternal aunt Hester, who was a wholehearted devotee of the non-juror's 'religious phrenzy'. She made an ostentatious show of her superior piety within his holy circle, becoming a recluse and even refusing to cross Gibbon's threshold, lest she be polluted. Thrifty with her own riches, she seems a prototype for Gibbon's swipes against the worldly failings of saints.

Gibbon thus grew up in an atmosphere charged with religious electricity, and had the run of a succession of private libraries stacked with works of controversial theology and divinity. Sacred history prompted the same transports as the *Arabian Nights*: 'my imagination was enchanted with the perpetual mixture of supernatural and human agents.' Its conundrums enthralled him: 'my sleep has been disturbed by the difficulty of reconciling the Septuagint with the Hebrew computation'; and, while yet a child, he tried to reconcile Scripture with the fabulous histories of Egypt and Babylon exactly in the manner of the times, as examined above in the first chapter: though Gibbon adds, with all due humility,

> at a riper age I no longer presume to connect the Greek, the Jewish, and the Egyptian antiquities which are lost in a distant cloud; nor is this the only instance, in which the belief and knowledge of the child are superseded by the more rational ignorance of the man.

Thus Gibbon's personal religious development recapitulated the religious history of the race and nation. 'From my childhood,' he confessed, 'I had been fond of religious disputation.' One of the earliest anecdotes about him reveals a precocity perilously primed with casuistry at its most sinister. He threatened (it was reported) to murder his beloved aunt Kitty. Why? she asked:

> He said, you are a very good woman and if you die now, may go to heaven; but if you live longer you may grow wicked and go to Hell – she remonstrated that he would certainly incur great guilt by such an action & be punished for it; & go where he wished to save her from going. He instantly replied, my Godfather will answer for it – as I am not confirmed.

The mature Gibbon well knew that theology could lead, through the most rational logic, to appalling deeds committed for the best of all possible reasons.

Gibbon thus grew up in the midst of religious ferment, reading 'High Church Divinity and much trash of a previous age'. While he was still young, two of his cousins converted to Catholicism; and he did so too, initially as a result of reading Middleton's *A*

Free Inquiry into the Miraculous Powers which are supposed to have subsisted in the Christian Church from the Earliest Ages through several successive Centuries (1749). The *Memoirs* tell us that, having gone on to read Bishop Bossuet's *Exposition of the Catholic Doctrine* (1671) and his *History of the Protestant Variations* (1688) he felt committed not just to an intellectual creed, however, but to all the trappings besides:

> The marvellous tales, which are so boldly attested by the Basils and Chrysostoms, the Austins and Jeroms, compelled me to embrace the superior merits of Celibacy, the institution of the monastic life, the use of the sign of the cross, of holy oil, and even of images, the invocation of Saints, the worship of relicks, the rudiments of purgatory in prayers for the dead, and the tremendous mystery of the sacrifice of the body and blood of Christ, which insensibly swelled into the prodigy of Transubstantiation.

Thus he became intimately acquainted with all he would later dub 'superstition'. The outcome of this 'momentary glow of Enthusiasm' was a bombshell: five years of exile in Lausanne, where he arrived 'a thin little figure with a large head, disputing and urging, with the greatest ability, all the best arguments that had ever been used in favour of popery'. Gibbon might well have remained loyal to his new faith, at the price of disinheritance and possibly exile. But he did undergo reconversion, though it was hard won and took eighteen months to achieve, finally concluding that Catholicism was contradicted by the best evidences of reason and the senses. There is no reason to believe that his initial return to the Protestant fold was anything but sincere. He did not leap 'from superstition to scepticism'; and only later did Gibbon abandon rational Protestantism and grow distanced, ironic and hostile, a Protestant only in Bayle's residual sense: '*je suis protestant, car je proteste contre toutes les religions.*' Some scholars have suggested that, despite the tone of his writings, Gibbon remained a Christian all his life (according to Parkinson, 'the *Decline and Fall* was written by a Protestant who was also a humanitarian in belief and a rationalist in method', and Turnbull has suggested that he probably nurtured a private piety). But, despite his routine church attendance, these views are not easy to reconcile with Gibbon's habitual animus against the Christian Churches and their destructive role in history.

This feverish period of conversion and reconversion left Gibbon feeling towards Christianity much as he came to conclude – though less strongly – about women: once bitten, twice shy. The potential magic of religion, as maybe of women, was profound, the dangers great. Prudence dictated that both were better resisted. After his Catholic affair and his passion for Suzanne Curchod, Gibbon never again compromised his independence. Hence his acute apprehensions of the seductive snares of Catholicism; visiting the Swiss monastery of Einsiedeln in 1756, he writes that it is 'proof of the potent magic of religion', at once 'a triumph of superstition, a masterpiece of ecclesiastical scheming, and a disgrace to mankind'; about the same time he remarks that certain prophecies must be taken as 'sorry evidence of the political craft of the powerful, the gullibility of the populace, the adulation of historians and the fraud of priests'.

Gibbon's one faithful love was for Rome, not the Roman Church. As 'the historian of the Roman Empire', he deplored the rise of Christendom because it had speeded the demoralization of the commonwealth he admired. In that most celebrated formula, the decline of the Roman Empire was the story of the 'triumph of barbarism and religion'. Concentrating upon inner processes, Montesquieu had been able to diagnose Rome's decease without mentioning religion; but Gibbon grew preoccupied with religion's power to make or mar nations, by providing social cement on the one hand, or by destabilizing and dividing on the other: 'For the man who can raise himself above the prejudices of party and sect, the history of Religions is the most interesting part of the History of the human Spirit.'

Gibbon may be seen as taking up the argument about the relations between religion and Rome first broached by St Augustine in his *De Civitate Dei* (*City of God*) and by his successor Salvian. For Augustine, writing soon after the sack of Rome by Alaric in 410, there was an urgent need to counter those old pagans who, long before Gibbon, were already blaming the Church for Rome's fall. Within a century of Constantine's conversion, the eternal city was no more. Why?

Augustine craftily deflected the accusation. It was not Rome which was the eternal city: that was the City of God. Precisely because it was of this world, part of the '*civitas terrena*' (earthly city), Rome had to perish; furthermore, being vainglorious, militaristic and still pagan, could it deserve to survive? The fate of early cities mattered little compared with 'God's City', the communion of the faithful destined for spiritual immortality. Because Christ's kingdom was not of this world, Christians must cultivate unworldliness. In Rome's fall lay no occasion for tears; in place of the Roman Empire, a new empire, God's, would emerge, a spiritual kingdom. The younger Salvian added a more aggressive coda. Rome was evil, Rome had persecuted the godly; now the Goths had been God's scourge.

Gibbon offered his own gloss to these ancient apologetics. The historian could not but agree with theologians like Salvian, that God was responsible for the fall of Rome. For that very unworldliness urged by Church leaders such as Tertullian and Lactantius had drained the political energies vital to a flourishing state. Thus Gibbon turned the theologians' arguments to his own purposes.

Gibbon regarded Christianity's cardinal virtues as indeed inimical to a commonwealth. Telling witness from the early Christians themselves revealed that they despised the political life, evaded civic responsibilities, and scorned secular culture, arts and learning: 'what has Athens to do with Jerusalem?' Christians encouraged believers to withdraw into preoccupation with personal salvation, rather than bolster civic welfare. Christianity had at best been indifferent to Rome's fate; at worst, it displayed passive hostility to the empire which, by and large, had shown it exemplary tolerance.

Similarly, Gibbon wholeheartedly accepted, though not in the same spirit, Augustine's other tenet, that a Christian empire would rise upon the ruins of Rome. For Augustine this constituted a prophet fulfilment of the divine plan. For Gibbon it had come to pass as a historical fact, which, like any other, the historian had to explain through a tracing of causal patterns, grounded upon a knowledge of men and affairs. Clearly Christianity had scored an extraordinary success. At first only one of a multitude of obscure sects in the Eastern Mediterranean, it had risen within three centuries to become the official faith of the empire; and 'on the ruins of the Capitol' it had 'erected the triumphant banner of the Cross'. Thereafter, in the West, 'Rome' meant the Roman Catholic Church and the Papacy (for Protestants, the Whore of Babylon) no less than the city of Romulus. Gibbon thus set out as a philosophical historian to explain – some might say, explain away – the human and natural causes of the rise of Christianity.

Herein lay one of Gibbon's crucial breaks with traditional sacred history. Catholics and Protestants had of course presented radically contradictory accounts of the early Church. But both had started from the premise that the line from the Old Testament patriarchs to the triumph of Christianity plotted the will of God, operating through Providence. The job of the Christian historian had thus been to trace the path of the divine plan, for which the Bible offered the best map. As Apthorp, one of Gibbon's own critics, put it, 'the sacred scriptures are so constructed, as to comprehend the general history of mankind, either in narration, or in prophecy.'

Gibbon never directly challenged this interpretation of the history of Christianity as God's will; he wrote on occasion, albeit tongue-in-cheek, about the suspension of 'the laws of nature ... for the benefit of the Church'. With tactical reticence he merely stated that his self-imposed task was the examination of 'the human causes of the progress and establishment of Christianity':

> Our curiosity is naturally prompted to inquire by what means the Christian faith obtained so remarkable a victory over the established religions of the earth. To this inquiry an obvious but satisfactory answer may be returned; that it was owing to the convincing evidence of the doctrine itself, and to the ruling Providence of its great Author. But as truth and reason seldom find so favourable a reception in the world, and as the wisdom of Providence frequently condescends to use the passions of the human heart, and the general circumstances of mankind, as instruments to execute its purpose, we may still be permitted, though with becoming submission, to ask, not indeed what were the first, but what were the secondary causes of the rapid growth of the Christian church?

With this programme Gibbon was proclaiming, rather as Voltaire had earlier in his article 'Histoire' in the *Encyclopédie*, the universal empire not of sacred but of secular history. Gibbon's detailed investigation did break new ground. There were as many histories of the Church as there were histories of Rome, but the investigation of their interrelation from the secular point of view had hardly been attempted; therein lay Gibbon's startling originality.

Yet Gibbon could draw upon a tradition, albeit a flawed one. For finding 'natural causes' for the rise of the Papacy was precisely the endeavour of Protestant scholars ever since the Reformation.

Reformers had elucidated that rise as the consequence of worldly ambition, greed and chicanery. They had portrayed power-hungry papists forging documents, manipulating princes with promises of power and heavenly reward, and exploiting popular gullibility through bogus miracles. Popes had craved the worldly power they hypocritically denounced.

Although Gibbon absorbed this 'radical Protestant' analysis, he did not resort to a crude 'conspiracy theory'; the early churchmen were not simply the Devil's disciples. In a properly historical way Gibbon was concerned, not so much to indict these turbulent priests, as to explain their success. He also carried his story back long before the political popes and their henchmen; like the Deists, he probed the early affairs of the Church as well. In a Protestant country this was the more daring act. In broadening his inquiry, Gibbon knew of course that he risked suggesting that the psychopathology of power which Protestants traditionally applied to Popery could also apply to the rise of the faith in general.

In other words, Gibbon the philosophical historian made religious belief itself into a historical problem. Why were there faiths at all? What were the human and natural springs of religious sensibility? As he suggested in his programme for examining secondary causes, Gibbon located such phenomena in 'the passions of the human heart and the general circumstances of mankind'. And to tackle this problem, he drew heavily upon a speculative psycho-anthropology of religion which had been first suggested by the writers of Antiquity, not least Herodotus and Lucretius, and elaborated by the *philosophes*, notably Fontenelle in his *Histoire des oracles* (1687) and Hume in his 'Essay on Superstition and Enthusiasm'.

From a philosophical point of view, this thesis ran, religion arose out of the ignorance, wonder and fear experienced by powerless primitive man confronted by a dangerous environment and a mysterious cosmos. Each object of terror – torrents, thunder, darkness, etc. – duly became an object of worship, abasement and placation, indeed a sacred being. Thus propitiated, objects were fetishized into gods. In other words, primitive religion was a defence mechanism; it generated beliefs which were essentially polytheistic, pluralistic and 'superstitious', in the sense of turning objects of sense experience into anthropomorphic objects of devotion, and personifying the forces of nature. Like other *philosophes*, Gibbon contended that the Christian Churches still contained vestiges of 'the dark and implacable genius of superstition', often manipulated by priests to exploit mass credulity. Obviously, Catholicism was particularly tainted, for Protestantism had specifically constituted a protest against popish superstition. The invocation of saints was polytheism in a Christian guise; prayer, pilgrimages and the 'sacrifice' of the Eucharist were latter-day propitiations; transubstantiation was magic.

In this analysis, superstition, though intellectually atavistic, was in one respect harmless to civil society. Grounded upon fear, it helped to underwrite political order: the superstitious were obedient. The peoples of the Roman Empire had evolved a great variety of superstitious cults, which could coexist harmoniously, because such primitive polytheism was inherently pluralistic. Thus the Romans could manifest a wise and salutary tolerance towards the plethora of cults within their jurisdiction. Tolerance was facilitated, not least because such cults lacked a professional priesthood and a church hierarchy: 'the ministers of Polytheism ... were, for the most part, men of a noble birth and of an affluent fortune. ... Confined to their respective temples and cities, they remained without any connection of discipline or government.' Religion had thus been compatible with the public good:

> The policy of the emperors and the senate, as far as it concerned religion, was happily seconded by the reflections of the enlightened, and by the habits of the superstitious, part of their subjects. The various modes of worship which prevailed in the Roman world were all considered by the people as equally true; by the philosopher as equally false; and by the magistrate as equally useful. And thus toleration produced not only mutual indulgence, but even religious concord.

Gibbon scarcely admired the picturesque absurdities of pagan polytheism; its saving grace, however, was that its creeds did not demand or command exclusive control over the minds of men. In a different respect, however, superstition had its dangers. For, by encouraging servility and credulity, it sapped the vitals of society, and prepared minds for tyranny.

For Gibbon and his philosophical mentors, religious experience underwent modification, like all other aspects of consciousness, in the general march of mind and society. Primitive polytheism could not rest in the face of the progress of knowledge, reason and action.

As increasing knowledge of nature became codified in science, religion itself had moved towards monotheism, postulating a single, unifying principle or person ordering what was increasingly perceived as a coherent universe. Mankind outgrew the mental degradation of primitive polytheism: 'As nations become more enlightened, so their idolatry becomes more refined. They understand more how the universe is governed closer to the unity of one efficient cause.' But a fresh danger sprang up in its stead, born of certainty and the new pride of mind: enthusiasm. The principle of one God seemed to require one faith, one truth, and even one chosen people. As the new monotheism supplanted the easy-going pluralist tolerance, an obsessional exclusivity was born, and with it unlimited scope for 'fanaticism' – 'one of those epidemic diseases of the human Spirit which merit great attention'. Monotheism liberated the mind from superstition, only to enslave it to spiritual despotism.

Potentially dangerous monotheistic sects had emerged during the classical period in the Eastern Mediterranean, associated with the philosophical religions of the Greek world and the rationalism of the earlier Ionians. Among the Greeks they remained small-scale, largely speculative, and essentially harmless. Monotheism also appeared, however, within Judaism – for which Gibbon felt that disgust shared by most *philosophes*. Jewish messianism was itself innocuous, because the Hebrews, though flattering themselves that they were the Chosen People, constituted an obscure, inward-looking and geographically localized group; that situation changed, however, when Judaism proved the midwife of Christianity. Then the full explosive potential of fanaticism was revealed.

For Gibbon, 'fanaticism' was more rational than 'superstition'. Hingeing upon a system of intellectual truth, its tendency was to support freedom, for it demanded the emancipation of the spirit. Yet such freedom for believers could readily spill over into the extremes of anarchy and subversion on the one hand, and intolerance on the other. Not least, the fanatic mind was easily enslaved to its own metaphysical fantasies, as Gibbon had already suggested in his controversy with Richard Hurd over the prophecies of the Book of Daniel, in which he diagnosed the prophetic strain as the disorder of reason pressing beyond its proper limitations:

The eager trembling curiosity of mankind has ever wished to penetrate into futurity; nor is there perhaps any country, where enthusiasm and knavery have not pretended to satisfy this anxious craving of the human heart ... These self-inspired prophets have strove by various arts to supply the want of a divine mission.

Drawing upon this framework for interpreting religion, Gibbon set about explaining the triumph of Christianity – though never its origin: he always spoke respectfully of Jesus as a man who, like Socrates, 'lived and died for the service of mankind', and, either for tactical or for personal reasons, never extended his history back to Gospel times. Amongst the multitude of late-classical faiths, it was Christianity which had become the official religion of the empire, and one of the agents in its fall. Why had it proved so strong? Gibbon singled out five features special to Christian monotheism which could explain this success.

The first was the impetus Christianity derived from zealotry itself, that inflexibility and intolerance it had inherited from the Jews (those 'enemies of the human race'). Although Judaism had been an exclusive cult, disinclined to proselytize, the Children of God were utterly 'obstinate' in their faith; the vision of the absolute divide between the godly and the ungodly had given Judaism its enormous resilience, and its unique sense of righteousness and divine protection. The early Christian Church absorbed these elements, but eagerly looked outwards with a mission to save mankind.

The second was the doctrine of the immortality of the soul and, above all, the promise of future bliss in heaven. As Warburton had stressed, this had not been a tangible tenet of Judaism; rather, the doctrine of the immortal soul and heavenly reward had emerged out of the philosophical speculations of the Greeks, especially the Platonists. Christianity appropriated and harnessed them to an explicit eschatology. God had created the world in time. He would likewise destroy it in time; indeed, most early Christians believed, in an imminent millennium. At Judgment Day, true believers would be rewarded and Satan's children damned. This 'advantageous offer' of blissful immortality boosted conversions and stiffened the morale of the faithful.

Third, Gibbon contended, claims to miracles and visions seemed to establish Christianity's truth and efficacy. Christ had raised the dead; the apostles healed by laying on of hands and spoke in tongues. Few of the old pagan religions had claimed such direct cosmic potency or thaumaturgic powers. This miracle-working magic – or at least its repute – had great appeal for the 'dark enthusiasm of the vulgar.'

Fourth, Christians commanded respect because of their superior conduct, austere morals and rigid rectitude. They offered themselves as models of holiness: pure in morals, proof against carnal temptations. Their penitence, asceticim and self-lacerations made powerful impressions. Under persecution they displayed exemplary fortitude; their ardour and zeal were unmatched: they were clearly worth emulating.

Last, Gibbon contended that Christianity made its way in the world because it organized itself so effectively. It set up a cellular network of churches; it appointed superior officials, bishops, who established what was initially a *de facto* system of subordination and unity; monastic communities took root. All these self-governing institutions, Gibbon noted perceptively, had the resilience and growth-potential possessed by comparable small political republics; and the example of early Rome may have been in his mind. Unlike the disorganized pagans, Christians drilled themselves into coherence: 'the union and discipline of the Christian republic ... gradually formed an independent and increasing state in the heart of the Roman Empire'. Gibbon attached pre-eminent importance to this factor, not least because it led to a distinct priesthood:

> The progress of the ecclesiastical authority gave birth to the memorable distinction of the laity and of the clergy, which had been unknown to the Greeks and Romans ... the latter [constitute] ... a celebrated order of men which has furnished the most important though not always the most edifying subjects for modern history.

Gibbon parades these causes before the reader with some detachment, not to say cynicism. He is of course concerned to establish neither the theological truth of Christianity's claims, nor even the sincerity of believers, but the impact achieved by their *appearances*; and it is perfectly consistent with his explanatory programme that occasional ironical shrugs should cast doubt on what underlay this show of faith. Thus Gibbon's discussions of miracles evince scepticism about their authenticity, while admitting their hold over primitive minds. Likewise, while evaluating the impact of the Christians' reputation for high moral standards, he slips in the counter-evidence of rapacity, deception and ambition. The unworldly, indeed other-worldly Christian doctrines somehow maximize their early powers; they profess moral austerity, yet Gibbon displays them time and again seizing the main chance. Qualities such as zeal, so crucial to the Christian success, countermand true humanity: 'The condemnation of the wisest and most virtuous of the pagans, on account of their ignorance or disbelief of the divine truth, seems to offend the reason and the humanity of the present age.'

As critics have always complained, Gibbon showed no sympathy for the Church's aspirations and shed no tears over its martyrs. This does not, however, vitiate his understanding or interpretation of the human causes of the rise of Christianity, in particular his thesis that the convergence of the factors he analysed granted the early Christians both tenacity and a dynamic capacity for expansion. Gibbon did not, however, exaggerate its appeal. Christianity did not spontaneously spread like wildfire and become the faith of the masses overnight:

> In the beginning of the fourth century the Christians still bore a very inadequate proportion of the inhabitants of the empire; but among a degenerate people, who viewed the change of masters with the indifference of slaves the spirit and union of a religious party might assist the popular leader to whose service, from a principle of conscience, they had devoted their lives and fortunes.

Christianity, long a minority sect, was nevertheless capable of perpetuating and defending itself because, unlike other cults, it evolved a powerful self-governing internal network.

By consequence, what was truly decisive to the fortunes of Christianity was the conversion of Constantine, the 'change in the national religion' as Gibbon put it. By ending persecution and encouraging the propagation of the faith, Constantine ensured the triumph of spirituality. Now with the blessing of the purple, Christianity was able to spread rapidly, on account of its zeal, its religio-ethical universalism, and its chain of command: 'the institution was so well suited to private ambitions and to public interest, that in the space of a few years, it was received throughout the whole empire.'

Thereafter all the absolutist tendencies latent within a fanatical religion remained unchecked by the civil authority, and indeed were often supported by it. Constantine unleashed a tiger. It is a gem of irony that Gibbon has so organized his narrative that the chapter (21) after his account of Constantine's politique conversion immediately launches into the proliferation of heresy and schism, and the adoption of persecution by the Church itself. Above all we see religious fanaticism on a scale hitherto unknown, rampant among the Donatists in Africa. Convinced of their own righteousness in an otherwise sinful world (they 'boldly excommunicated the rest of mankind'), these heretics turned themselves into brigands, creating carnage in the name of love and charity, and ultimately speeding the loss of North Africa to the Vandals.

Theological controversy also began to unsettle the empire. Incorporating the metaphysical speculations of Platonism, Christianity enveloped itself in systems of rational theology ('a strange centaur!' Gibbon elsewhere remarks). The mystery of the Trinity was among the first to be elaborated, though three rival renderings of the ultimate unity of the Godhead – Arianism, Tritheism and Sabellianism – were soon jockeying for supremacy. During the age of the great Councils (Nicaea, Ephesus, Chalcedon, etc.), theological hairsplitting, Gibbon suggests, brought the empire almost to a standstill. In this 'dark abyss of metaphysics', the statesmanship of Athanasius, besieged by the implacable bigotry of the righteous, shone out alone.

Thereafter, the history of Christianity was the history of heresy, schism and fraternal persecution. Such battles represented dogmatism run riot on matters not properly open to the testimony of the senses, as with the long-running controversy over the relations between the Persons of the Trinity which split East and West, affording Gibbon some sport at the expense of this perversion of a faith which affirmed that truth came out of the mouths of babes and sucklings:

> In the treaty between the two nations several forms of consent were proposed, such as might satisfy the Latins without dishonouring the Greeks; and they weighed the scruples of words and syllables, till the theological balance trembled with a slight preponderance in favour of the Vatican. It was agreed (I must intreat the attention of the reader), that the Holy Ghost proceeds from the Father *and* the Son, as from one principle and one

substance; that he proceeds *by* the Son, being of the same nature and substance; and that he proceeds from the Father *and* the Son, by one *spiration* and production.

Gibbon emphasizes three elements of the faith which had established itself by the onset of the Dark Ages. He was, first, both intrigued and appalled by the Christian repudiation of all that had been the glory of the commonwealth. No one could be more unlike the Roman citizen than the Christian monk; monasticism and political life were antitheses: 'The lives of the primitive monks were consumed in penance and solitude, undisturbed by the various occupations which fill the time, and exercise the faculties of reasonable, active, and social beings.' Gibbon sees asceticism and its corollary, the theology of radical sinfulness, setting profound historical puzzles; for both seemed to have an inherent tendency to run to paradox, a fact noted by Wesley no less than Gibbon. Vows of chastity created obsessions with sexuality and led virgins into compromising situations; monastic poverty resulted in extremely wealthy monasteries (Gibbon never forgot Einsiedeln). Gibbon enjoys the irony while, of course, using it to score a more profound point: these unintended and self-defeating consequences arose out of the intrinsic unnaturalness of Christian asceticism. It is a degrading philosophy which robs man of all that dignifies him. Furthermore, its denial of human nature inevitably leads to perversion. It anaesthetizes our sensibilities towards our fellow men: mortification is the double of fanaticism: he who hates himself, hates others. And, most terrifyingly of all, the solitary mode of life adopted by ascetics releases them from all the normal constraints which the web of society naturally imposes upon social animals; hermits waft off into solipsistic fantasies of thought and action, and irretrievably abandon themselves to their own imaginations. Fanaticism, in servitude to the tyrant God, is the antithesis of life. One should never forget Gibbon's capacity for simple joy: 'next wednesday I conclude my forty-fifth year. ... I am very glad that I was born.'

Second, Gibbon turns his gaze to the dogmatic fanaticism of Christian theology itself: the doctrinal wars over the Trinity, the divinity of Christ, the use of images and the worship of saints. Gibbon's attack hinges on Lockean empiricism. These disputes cannot be resolved by the evidence of our senses; they are quarrels about ghosts and figments, and even, literally, about iotas, as in the fourth-century debate over the divine essence:

> The Greek word which was chosen to express this mysterious resemblance bears so close an affinity to the orthodox symbol, that the profane of every age have derided the furious contests which the difference of a single diphthong existed between the Homoousians [united in essence] and the Homoiousians [similar in essence].

Such hairsplittings are absurd: 'these speculations, instead of being treated as the amusement of a vacant hour, became the most serious business of the present, and the most useful preparation for a future life.' The only certain path, Gibbon suggests, is to follow Procopius in a confession of ignorance about the true nature of the one God whom reason leads us to revere. This fideistic formula leaves it unnecessary to ask if, in this, he was being sincere.

Philosophical debates are harmless, but theological disputations are pernicious nonsense. Expounded in a spirit of doctrinaire fanaticism, theology established its tyranny over the intellect. The creed of one God, one faith and one truth sanctioned the Inquisition, the Index and the stake.

Gibbon retained a personal horror of the snares of theological dogmatism into which it had proved all too easy for his logic-chopping brain to fall. The danger, he knew, had hardly diminished over the centuries. As Pocock has rightly insisted, Gibbon was not disposed to view Protestantism as intrinsically more liberal or tolerant than Catholicism. Fideistic Catholicism accepted that many articles of faith rested merely on tradition or authority, but Protestantism's commitment to theological rationalism was especially perilous.

Gibbon regarded the wrangling Dissenters of his own times as the ghosts of the word-spinning nonsense-grinders of ancient Hellenistic neo-Platonism. He daringly traced the history of rational dissent as a progression from the Gnostics to the Paulicians to the Cathars and then on through the early Protestants ('a crowd of daring fanatics'), and finally to the Arians of his own day. Gibbon despaired of Calvin's vengeful God ('many a sober Christian would rather admit that a wafer is God than that God is a cruel and capricious tyrant'), and the zeal of the likes of William Law – 'hell-fire and eternal damnation darted from every page of the book.' But what terrified him above all was the entrenched righteousness of the rational Protestant, as typified by his contemporary, the radical Dissenter Joseph Priestley.

Challenged to a theological 'duel' by Priestley, Gibbon felt justified in counter-accusing his Arian antagonist – whom he pictured as a kind of leftover Church Father, 'outstripping the zeal of the Mufti and the Lama' – of fostering unbelief itself by his tortuous, restlessly intellectualizing theology:

> That public will decide to whom the invidious name of *unbeliever* most justly belongs; to the historian, who, without interposing his own sentiments, has delivered a simple narrative of authentic facts, or to the disputant, who proudly rejects all natural proof of the immortality of the soul, overthrows (by circumscribing) the inspiration of the Evangelists and Apostles, and condemns the religion of every Christian nation as a fable less innocent, but not less absurd, than Mahomet's journey to the third Heaven.

Gibbon reprobated religious bigotry, both Catholic and Protestant, ancient and modern. Thus he finds it deplorable that the 'vigorous mind' of Samuel Johnson was 'greedy of every pretence to hate and persecute those who dissent from his creed'. But one did not have to be a *dévot* to be a bigot. Gibbon saw a new epidemic breaking out in the smart salons of Paris, led by Voltaire's '*écrasez l'infâme*'. He strategically contrasted Hume's humanity with Voltaire's zeal; Voltaire 'laughed at the scepticism of Hume, preached the tenets of atheism with the bigotry of dogmatists, and damned all believers with ridicule and contempt'. The doctrinaire spirit was hardly capable of reform. Calvin was just as bigoted as Loyola, and felt no compunction about burning Servetus – a man already condemned by the Catholic Church – in the name of the true doctrine of the Trinity. Eventually, of course, Protestantism had become associated with greater tolerance, but this arose not through any virtue of the Reformers but merely because tolerance was forced upon them in enlightened nations such as England and Holland by the civil authorities: 'The nature of the tyger was the same, but he was gradually deprived of his teeth and fangs.'

As Gibbon saw it, bigotry was an expression of hate; Christians had complained about persecution in the early days, but 'the primitive church ... delivered over, without hesitation, to eternal torture, the far greater part of the human species.' Putting truth before people brought division and enmity wherever it went. When rival sects faced each other in the name of one God and one faith, schism became endemic. Gibbon traced with great skill the splitting off and subsequent expulsion of the Nestorian, the Coptic, Abyssinian, Monophysite, Syrian and Maronite Churches and so forth:

> Under the rod of persecution, the Nestorians and Monophysites degenerated into rebels and fugitives; and the most ancient and useful allies of Rome were taught to consider the emperor not as the chief, but as the enemy of the Christians.

Each damned the others, and no matter how small – or precisely *because* they were so small – believed itself uniquely godly. 'Even the imperceptible sect of the Rogatians could affirm without a blush, that when Christ should descend to judge the earth, he would find his true religion preserved only in a few nameless villages of the Caesarean Mauretania.' Theological mania eventually led to fundamental splits between the Western and Eastern Churches over the use of images (the iconoclastic controversy of the seventh and eighth centuries), followed by the formal rupture between Latin Christendom and Greek Orthodoxy in the eleventh century. Provocatively for one writing in Protestant England, Gibbon reduced the Reformation itself to a schismatic ripple.

Christianity divided. But worse – and this was the third main feature Gibbon stressed – it had also come with a sword. Gibbon takes wicked relish in retailing the bloodcurdling tales of mob violence amongst the thugs of the early Church, in particular the gang warfare routinely disgracing Church politics at Alexandria and Constantinople, where there was internecine struggle between the Green Party and the Blues. '"May those who divide Christ be divided with the sword, may they be hewn in pieces, may they be burned alive," were the charitable wishes of a Christian synod.' Gibbon is pleased to show how monastics were at the thick of the evil. At one stage 'Jerusalem was occupied by an army of monks; in the name of the one incarnate nature, they pillaged, they burnt, they murdered; the sepulchre of Christ was defiled with blood.' The persecution of Christians by pagan emperors was never so bloody as the subsequent persecution of heretics and schismatics by Christians, once Constantine had pledged to support the zeal of the Church with the sword of the state. Justinian, for instance, 'piously laboured to establish with fire and sword the unity of the Christian faith', but his labours were counter-productive for the empire. His attempts to enforce orthodoxy in Egypt merely meant that Islam was all the more readily received.

Nor was the persecuting spirit simply a growing pain of the early Church, insecure and attempting to establish itself. Indeed, in ecclesiastical matters too, power had corrupted, and Gibbon traces the career of this ferocious spirit right into the modern world. Far more Christians were murdered by Christians during the Reformation than the tally of believers sacrificed in three centuries of Roman persecution, when there was merely, at his computation, 'an annual consumption of one hundred and fifty martyrs'.

Christianity thus stands indicted as destroying civilization. No wonder 'that the propagation of the gospel and triumph of the Church are inseparably connected with the decline of the Roman Monarchy'. Gibbon summed up his judgment in his 'General Observations on the Fall of the Roman Empire in the West':

> The clergy successfully preached the doctrines of patience and pusillanimity; the active virtues of society were discouraged; and the last remains of military spirit were buried in the cloyster: a large portion of public and private wealth was consecrated to the specious demands of charity and devotion; and the soldiers' pay was lavished on the useless multitudes of both sexes, who could only plead the merits of abstinence and chastity. Faith, zeal, curiosity, and the more earthly passions of malice and ambition, kindled the flame of theological discord; the church, and even the state, were distracted by religious factions, whose conflicts were sometimes bloody, and always implacable; the attention of the emperors was diverted from camps to synods; the Roman world was oppressed by a new species of tyranny; and the persecuted sects became the secret enemies of their country.

Gibbon offers us the anatomy not just of two major belief-systems – paganism and Christianity – but of three. For in his fifth volume Islam takes to the stage. The question of how precisely to interpret the nature and rise of Islam had become, particularly from the seventeenth century, vital to European religious, political and historical debate, for it had obvious and immense ramifications for the relations between Church and state, clergy and laity, faith and reason, and for the whole explanation of religiosity. In the polemical world of the early Enlightenment, renderings of Islam – favourable or unfavourable – were standardly read as covert analyses of the Christian Church itself.

One interpretation of Islam sought to expose it as a false faith. It could be represented as a religion founded by a fraud, superstitious and ambitious, passing himself off as a divinely inspired prophet; this was essentially the line taken in Humphrey Prideaux's *The True Nature of Imposture Fully Displayed in the Life of Mahomet* (1697), a work written by a pillar of Anglican orthodoxy, who meant it to serve as a public warning against the pretensions of radicals and reformers. This interpretation drew attention to the footling tales and trumped-up miracles presented in the myths arising around Mohammed; Islam was clearly a faith founded upon credulity. The teachings of the Prophet moreover embraced a convenient self-serving immorality, e.g. through sanctioning polygamy. Not least, Mohammed had turned to fanaticism. Religion became the specious pretext for conquest and slaughter.

Prideaux's account was intended to discredit Islam through pointing up its *differences* from the true faith. Clearly, such an analysis could be deployed for the opposite purpose, by hinting, not at difference, but at the deep similarities between the fortunes of this false, silly and depraved religion and those of Christianity itself. Indeed, there was an element of this approach even in Prideaux, as also in Charles Leslie's *Socinian Dialogues* (1708–10). For both writers pointed out that, theologically speaking, modern Unitarianism – which of course they deeply deplored – bore close resemblance to the theology of Islam. Committed to perceiving the finger of Providence even in evil, Prideaux was inclined to see Mohammed as divinely sent: 'He raised up the Saracens to be the Instruments of His Wrath.'

A completely different rendering of the rise of Islam was, however, possible, one advanced above all in the Count de Boulainvilliers' *Vie de Mahomed* (1730). Here Islam was represented as preferable to Christianity – tolerant, unmystical, undogmatic. Mohammed's theology had a becoming modesty. All he required to be believed was that there was one God ('a necessary truth,' commented Gibbon) and one prophet ('a necessary fiction'). Moreover, Mohammed set up no priesthood. Anticlerical Europeans professed their admiration for Islam for having revered wise men, learned in the sacred writing, while stopping short of a privileged, rich and mighty clerical estate. Indeed, so this interpretation ran, Islam did not sap the sinews of the state by preaching otherworldliness; rather it squared well with man's civil duties, by encouraging such values as honour, hospitality and justice. Those in Enlightenment Europe seeking models of what a priest-free and theology-free religion might resemble, found Islam a very attractive option.

Gibbon had had his interest in Islam roused at an early age:

> Muhammed and his Saracens soon fixed my attention, and some instinct of criticism directed me to the genuine sources. Simon Ockley, an original in every sense, first opened my eyes, and I was led from one book to another, till I had ranged around the circle of Oriental history.

Faced with his inability to read Arabic, and the polemical slant – indeed, politicization – of existing scholarship, Gibbon had difficulty in steering his interpretation of Islam on a sure course. This was partly due to historical scruples; the sources were poor and biased, and he had to confess that one could approach Mohammed himself only through 'a cloud of religious insense'. He did, however, lean towards the Boulainvilliers' interpretation. Historically speaking, Gibbon thought it was of cardinal importance that Islam did not develop a distinct priestly caste. He stressed how, in contrast to the tortuous elaborations of Christian theology, Islam retained its primordial simplicity – doubtless a function of the absence of a professional priesthood. For Gibbon, Islam activated rather than enervated society, effectively providing, as good civil religions should, an ideological bond. Moreover, being loosely organized, Islam never had the same power to hinder progress as the inquisitorial Christian Churches long possessed.

Yet like Christianity Islam came with the sword. Within a century of the Prophet's death, the Saracens had not merely conquered the Levant and the southern shores of the Mediterranean, but had advanced up through Spain into France as far as the Loire valley. Had it not been for Charles Martel's famous victory, Gibbon amused himself by speculating, we might today have the Moors installed in Magdalen: 'perhaps the interpretation of the Koran would now be taught in the schools of Oxford.' Indirectly, the rise of Islam then produced the Crusades; eventually, of course, Islam retraced the footsteps of the Christian Crusaders and planted the Koran and the Crescent in Constantinople:

While the [Byzantine] state was exhausted by the Persian war, and the church was distracted by the Nestorian and Monophysite sects, Mahomet, with the sword in one hand and the Koran in the other, erected his throne on the ruins of Christianity and of Rome.

In Gibbon's vision of the long centuries of decline, no civil polity could ultimately withstand the destructive forces of the Church Militant. Thus if the *Decline and Fall* opened with an era of civil peace, it concluded with centuries of religious warfare, mainly caused by Christians themselves: 'all that history has recorded [shows] that the Christians, in the course of their intestine dissensions, have inflicted far greater severities on each other, than they had experienced from the zeal of infidels.'

Gibbon's hostility to religion, and especially to Christianity, is often taken as a symptom of some personal immaturity, a character flaw, a spiritual deafness, an inability to surmount the scarring which his early conversion had caused. 'He often makes, when he cannot readily find, an occasion to insult our religion,' wrote Porson, 'which he hates so cordially that he might seem to revenge some personal injury.'

The result, it has been claimed, is that, when dealing with religion, Gibbon substituted the stilted rhetoric of the Enlightenment for true historical analysis, churning out just that kind of cliché history he elsewhere deplored; lacking a deeper grasp of the dynamics of faith, he resorted to irony. This view has its point. It is a somewhat damning indictment of Gibbon that he never steps far beyond the gut prejudices of *engagé* Protestant historians when dealing with the history of monks and nuns, popes and scholastics, nor beyond the banal stereotypes of Catholic historians when covering the rise of heresy and the schismatics. (I say not far, because Gibbon, almost despite himself, does in fact offer a handful of sympathetic portraits of individual churchmen, not least Athanasius and Ambrose.) In general, Christianity receives a kind word from Gibbon only when it produces unintended consequences such as helping to civilize the barbarians:

the pure and genuine influence of Christianity may be traced in its beneficial, though imperfect, effects on the Barbarian proselytes of the North. If the decline of the Roman empire was hastened by the conversion of Constantine, his victorious religion broke the violence of the fall, and mollified the ferocious temper of the conquerors.

We should not, however, be too judgmental. Like his brother *philosophes*, Gibbon could look back with unfeigned horror upon the fanaticism which had brought the sword and the Inquisition to Europe almost within living memory. Witches were still being burned in his lifetime. When Gibbon sought the roots of this bigotry, he did not find them in Cicero but in Patriarch Cyril of Alexandria – if there is a real devil in the *Decline and Fall*, it is he. Enthusiasm could easily be traced back almost to the very origins of Christianity; it could undoubtedly be traced back to religion in his own childhood and breast.

Gibbon was not so unsubtle as to offer a blanket condemnation of religion *per se*, and he saw some hopeful signs around him. He was amused for example by the fact that most Anglican clerics no longer treated the Articles of their confession very seriously, and doubtless he enjoyed the occasion when 'the historian of the Roman Empire' partnered the wife of the Archbishop of Canterbury at whist. What Gibbon sought to unmask were the evils of fanaticism. This was not an irrelevant, self-indulgent project, because such evils still surfaced. Above all, they exploded in the anti-Catholic Gordon Riots, which swept London in 1780, ending in hundreds of deaths and vast destruction of property, proof of the zealotry still aflame within Protestantism itself: 'Forty thousand Puritans, such as they might be in the time of Cromwell, have started out of their graves,' deplored Gibbon; it was 'the old story of religion' yet again. As he wrote to his step-mother:

> Our danger is at an end, but our disgrace will be lasting, and the month of June 1780 will ever be marked by a dark and diabolical fanaticism, which I had supposed to be extinct, but which actually subsists in Great Britain perhaps beyond any other Country in Europe.

Bigotry, though not in the name of religion in the strict sense, then of course broke out with greater ferocity in the new faith of the French Revolution, when Gibbon had cause to speak of 'the Jacobin Missionaries'. And these outbreaks of religious vandalism and barbarism merely echoed the spirit of the band of theological crusaders against *Decline and Fall*, whose assurance of infallibility matched their deep ignorance.

Gibbon's critics hurled brickbats and damned the author. But they had no better history of the rise of Christianity to put in its place. Indeed, well into the nineteenth century, Dean Milman and even Cardinal Newman could lament that Gibbon's still remained the most scholarly ecclesiastical history available: 'It is melancholy to say it,' wrote Newman, 'but the chief, perhaps the only, English writer who has any claim to be considered an ecclesiastical historian, is the unbeliever Gibbon.'

In the light of the furore his book caused, of the Gordon Riots, and not least of the Evangelical revival gathering pace, it would be silly to suggest that Gibbon was merely tilting at windmills when he investigated the continuing psychopathology of enthusiasm. It might be said that 'the historian of the Roman Empire' never truly understood the forces which fanned the flames of bigotry. The fact that the Reverend Thomas Bowdler brought out a fully 'bowdler-ized' version of the *Decline and Fall*, with the offensive religious chapters totally excised, suggests, however, that he understood them only too well.

6 Civilization, Barbarism and Progress

There is the moral of all human tales;
'Tis but the same rehearsal of the past,
First Freedom, and then Glory — when that fails,
Wealth, vice, corruption, — barbarism at last.
And History, with all her volumes vast,
Hath but *one* page ...
                    Byron, *Childe Harold's Pilgrimage*

At the close of the third volume of his epic, as he reached the end
of the eighth century, Gibbon offered his 'General Observations
on the Fall of the Roman Empire in the West'. Today's historians
find them disappointing: the analysis, it has been said, 'does not
take us very far'. This is principally because Gibbon opens by seemingly begging the question. Accounting for 'the rise of a city, which
swelled into an Empire' would indeed pose explanatory difficulties
and 'may deserve, as a singular prodigy, the reflection of a philosophic mind'. Not so, however, its *fall*, for:

> the decline of Rome was the natural and inevitable result of
> immoderate greatness. Prosperity ripened the principle of decay;
> the causes of destruction multiplied with the extent of conquest;
> and as soon as time or accident had removed the artificial sup-
> ports, the stupendous fabric yielded to the pressure of its own
> weight.

The decline of empires was not difficult to explain. 'The story of
its ruin is simple and obvious; and instead of inquiring *why* the
Roman empire was destroyed, we should rather be surprised that
it had subsisted for so long.' Gibbon did, however, make some
effort to satisfy the 'inquiry', albeit in uncharacteristically breathless
prose:

> The victorious legions, who, in distant wars, acquired the vices
> of strangers and mercenaries, first oppressed the freedom of the

republic, and afterwards violated the majesty of the purple. The emperors, anxious for their personal safety and the public peace, were reduced to the base expedient of corrupting the discipline which rendered them alike formidable to their sovereign and to the enemy; the vigour of the military government was relaxed and finally dissolved by the partial institutions of Constantine; and the Roman world was overwhelmed by a deluge of barbarians.

Examining this and other passages, Curtis concluded that Gibbon was 'strangely muddled' about the exact causes of Rome's fall, and many scholars have found his account banal.

This is perhaps not surprising, as Gibbon was here writing within a well-worn tradition. His vision of greatness sowing the seeds of its own destruction echoes Montesquieu, and above all the account presented by William Robertson in his *View of the State of Europe* (1769). 'There were defects in the Roman government,' judged Robertson:

even in its most perfect form, which threatened its dissolution. Time ripened these original seeds of corruption and gave birth to many new disorders. A constitution unsound and worn out must have fallen into pieces of itself without any external shock. The violent irruption of the Goths, Vandals, Huns, and other barbarians, hastened the event, and precipitated the downfall of the Empire.

The powerful dialectics of Montesquieu, Robertson and Gibbon, their insight that societies carry conflicting elements, the very structures of which will resolve themselves only after trouble and time, is a more sophisticated version of Renaissance notions of organic cycles of youth, maturity and the decay of old age, and the even older idea of fortune's wheel: what goes up must come down.

It is true that Gibbon's 'Observations' offer no pecking order of the distinct causes, proximate and ultimate, which conspired in Rome's downfall. Gibbon does not explain the relative importance of internal and external causes of 'barbarism' and 'religion'. But that is not his intention; he promised 'General Observations', not a causal analysis. The tabulation of a class-list of causes, or the unveiling of some grand arcanum, an ultimate hidden cause, would have suspiciously resembled plotting the pathways of Providence. Neither did Gibbon regard investigating the course of human affairs as analogous to a laboratory experiment – with empires falling like apples – in which pairings of causes-and-effects could be systematically isolated. As befitted an admirer of the sceptical Hume, Gibbon set little store by causal analysis anyway, finding consequences more significant. When he says that it is Rome's *endurance* which is remarkable, not its *fall*, we should take him at his word. Decay and corruption seemed natural and unproblematic to Augustan minds; and, as Peter Brown has so perceptively stressed, what particularly captured Gibbon's attention were the conditions under which such a highly precarious fabric of order, justice and mutual responsibilities which comprised that stupendous civilization could exist at all. As Gibbon insists, the larger the empire, the greater the contradictory tensions tugging at its fabric, those robes of civilization barely covering the inordinate passions of man.

Crucial to these 'General Observations' – and it is the very heart of Gibbon's vision – is the thesis that Europe's greatest civilization was conquered by barbarism. By contrast to the optimism of such *philosophes* as Turgot, Gibbonian history obeyed no inevitable plan of progress, natural or providential. Gibbon shed few tears at the fall of the empire as such, for he had long entertained the gravest doubts about all-conquering Rome. As he had written so emphatically to his father in 1764: 'I am convinced there never never existed such a nation, and I hope for the happiness of mankind that there never never will again.' If Rome as such unsettled him, imperial Rome troubled him still more deeply. For Gibbon, like so many of his contemporaries, identified political and military empires with misery. '*L'histoire des empires est celle de la misère des hommes,*' is the arresting opening of Gibbon's *Essai*, itself a secularized gloss upon Bossuet's pious '*les révolutions des empires sont réglées par la providence et servent à humilier les princes.*' Gibbon's horror of empire was of a kind with William Blake's, and he could have endorsed the ideas of Shelley, who, however, as a Romantic idealist, was predictably snooty about the sceptical historian.

It is, of course, true that Gibbon depicted the era of the early empire as the acme of happiness: 'if a man were called to fix the period in the history of the world, during which the condition of the human race was most happy and prosperous, he would, without hesitation, name that which elapsed from the death of Domitian to the accession of Commodus.' But Gibbon did not believe that such felicity was due to the intrinsic qualities of empire itself. And he knew that some readers would sense the sting in the tail of his judgment. For his apparent paean to empire was, *mutatis mutandis*, taken verbatim from William Robertson, who had noted in his view of the Roman Empire just three or four centuries later:

> if a man were called to fix upon the period in the history of the world during which the condition of the human race was most calamitous and affected, he would, without hesitation, name that which elapsed from the death of Theodosius the Great, to the establishment of the Lombards in Italy.

The point of Gibbon's ostentatious borrowing was precisely to underline the tragedy of empire: how decisively its progress had, in the course of just three centuries, subverted human well-being. The age of the Antonines was happy because it had inherited the virtue of the republic. But the old freedom of the Senate and people of Rome vanished like smoke – indeed even its *memory* was lost, or rather obliterated: 'it was artfully contrived by Augustus that, in the enjoyment of plenty, the Romans should lose the memory of freedom.' And, even then, perturbed that he had still painted the early empire in too favourable a light, Gibbon later regretted his decision – one probably taken largely to avoid treading in Tacitus' footsteps – to give the peaceful age of the Antonines such prominence at all: the evils of empire in the first century AD should have been made more conspicuous from the start. Contemplating a revised version, he asked:

> Should I not have deduced the decline of the Empire from the Civil Wars, that ensued after the fall of Nero or even from the tyranny which succeeded the reign of Augustus? Alas! I should; but of what avail is this tardy knowledge? Where error is irretrievable, repentance is useless.

Had Gibbon begun with Nero, no reader could have had any doubts about history's verdict upon the evils of empire. But even as it was, the major part of the first two volumes presented a vision of gradual immiseration and spiritual desolation, up to the point when, by the fifth century, 'if Rome still survived, she survived the loss of freedom, of virtue, and of honour.' It was well that new blood conquered:

> This diminutive stature of mankind ... was daily sinking below the old standard, and the Roman world was peopled by a race of pigmies, when the fierce giants of the north broke in, and mended the puny breed.

Gibbon radically distrusted empire, but he admired civilization, and saw Rome as one of its noblest edifices. Though civilization had its discontents, Gibbon was confident that Rome had conferred upon a large portion of mankind a unique blend of civic virtue and freedom, the blessings of order, justice, law, prosperity and individual rights, and, not least, a cultural heritage of poetry, oratory, history, art and moral philosophy. Extending civilization throughout the Mediterranean basin, Rome tempered her rule with justice and clemency, enhanced with religious tolerance and a remarkable freedom from racial chauvinism. Gibbon may have had mixed feelings about the Romans, but he never belittled civilization. Unlike many of the later *philosophes*, he never questioned that the life of society, of politics, arts and letters represented real achievement. He showed not an ounce of sympathy for, nor even interest in, the parlour primitivism of Rousseau. Nor was Gibbon's blood stirred by the expansion achieved by naked military power; the true art of the victor was to impose a peace beneficial to victors and vanquished alike. No democrat or populist, Gibbon's criterion for a polity was nevertheless the general welfare of the people: 'if we are most deeply affected by the ruin of a palace than by the conflagration of a cottage, our humanity must have formed a very erroneous estimate of the miseries of human life.'

The barbarian twilight began spreading over the empire in the late fourth century. The Goths crossed the Danube, eventually sacking Rome in 410. Germanic tribes flooded into Gaul; soon the Vandals seized North Africa, the Visigoths established themselves in Spain and southern Gaul and finally the Huns themselves, under Attila, swept across the Central European plain. The Western empire collapsed. The great wall, which so long had divided civilization and barbarism, was broken down:

> This memorable passage [of the Rhine] of the Suevi, the Vandals, the Alani, and the Burgundians, who never afterwards retreated,

may be considered as the fall of the Roman empire in the countries beyond the Alps; and the barriers, which had so long separated the savage and the civilized nations of the earth, were from that fatal moment levelled with the ground.

How did Gibbon interpret these barbarians? Reading the *Decline and Fall*, Horace Walpole – himself, of course, fascinated by the 'Gothic' – commented on the 'strange contrast between Roman and Gothic manners'. That was precisely the polarity which Gibbon chose to emphasize. Barbarism was the negation of civilization. All men had similar appetites and faculties; and of course vainglorious civilized man all too often acted barbarically and nullified the distinction: 'the same effects must be produced by the same passions; and when those passions may be indulged without control, small alas! is the difference between civilized and savage man.' Yet it had been the mission, and often enough the achievement, of Rome to promote higher standards of life over and against the almost un- or pre-historical world of the barbarians.

Gibbon built his idea of the Goths upon Tacitus' accounts of the Germanic peoples, and developed it in the light of those analyses of man's progress from rudeness to refinement set out by Scottish philosophers and political economists, above all by Adam Ferguson and that 'sage and friend' Adam Smith. These theorists argued that civil society had evolved through a universal series of stages from the primitive savagery of hunters and gatherers, through the 'barbarism' of nomadic pastoralists, to the feudal organization of agriculture, finally to emerge in commercial market society. Gibbon's barbarians were true nomads, and he stressed how the inevitable limitations of pastoralism fixed them on, or just beneath, the very bottom rung of the ladder of civilization. The Goths lacked so many of the essential attributes of civilized life, those qualities which were the starting-blocks of advance. They were, for one thing, 'unacquainted with the use of letters', the *sine qua non* of further progress:

[the] principal circumstance that distinguishes a civilized people from a herd of savages incapable of knowledge or reflection. Without that artificial help, the human memory soon dissipates or corrupts the ideas intrusted to her charge; and the nobler faculties of the mind, no longer supplied with models or with materials, gradually forget their powers.

In other words, being illiterate, the barbarians made few improvements and, what is more, lacked means to preserve and transmit them.

Moreover they had not yet developed fixed, landed property and the laws to regulate it – a crucial mark of primitivism, for 'the possession and enjoyment of property are the pledges which bind a civilised people to an improved country.' Nor had they invented money. Without land or a fixed medium of exchange, barbarians could never appreciate the cumulatively beneficial effects of labour, which, as Adam Smith argued, was the source of all wealth. Gibbon thus spelt out the intimate connexions between economic and social progress:

> In a civilised state, every faculty of man is expanded and exercised; and the great chain of mutual dependence connects and embraces the several members of society. The most numerous portion of it is employed in constant and useful labour. The select few, placed by fortune above that necessity, can, however, fill up their time by the pursuits of interest or glory, by the improvement of their estate or of their understanding, by the duties, the pleasures, and even the follies of social life. The Germans were not possessed of these varied resources.

What the barbarians did possess, however, were rude courage and a love of elemental freedom; and, of course, in a negative way, 'their poverty secured their freedom, since our desires and our possessions are the strongest fetters of despotism.' The irony of history, however, was that the Goths were unable to put that liberty to good use. For they had no settled societies, being essentially leaderless except when temporarily commanded by a prodigy such as Alaric. The ancient British tribes, commented Gibbon, characteristically 'possessed valour without conduct, and the love of freedom without the spirit of union': it was a recipe for defeat. Barbarians, fixed in the infancy of the race, displayed a childlike and perplexing mix of indolence and restlessness. As one who deplored the romanticization of childhood, Gibbon was nettled by those twaddling intellectuals who waxed lyrical about noble savages:

> They massacred their hostages, as well as their captives: two hundred young maidens were tortured with exquisite and unrelenting rage; their bodies were torn asunder by wild horses, or

their bones were crushed under the weight of rolling waggons; and their unburied limbs were abandoned on the public roads, as a prey to dogs and vultures. Such were those savage ancestors, whose imaginary virtues have sometimes excited the praise and envy of civilized ages.

It was crucial to Gibbon's reading of the encounter of Romans and Goths to stress the backwardness of the barbarians who 'passed their lives in a state of ignorance and poverty, which it has pleased some declaimers to dignify with the appellation of virtuous simplicity'. The struggle between Romans and Goths was not just a series of military trials between two warrior peoples. For Gibbon it marked a turning-point, indeed a turning-back, in the course of human society.

The primitiveness of the Goths became all the more conspicuous when, once installed within the old imperial boundaries, they took their first steps towards settled civilization. Lacking the mental fibre of civilized man, they tumbled from barbarism to corruption at one step. Almost as soon as they established themselves in Spain or Italy, they were seduced by luxury and enervation: 'Alaric would have blushed at the sight of his unworthy successor, sustaining on his head a diadem of pearls, encumbered with a flowing robe of gold and silken embroidery, and reclining on a litter or car of ivory, drawn by two white mules.' The fierce horsemen of the Arabian desert were corrupted the moment they tasted the fruits of civilization at Baghdad; and even, amongst the Mongols, the successors of Kubla Khan 'polluted the palace with a crowd of eunuchs, physicians and astrologers'.

In short, barbarism was a state hardly within the pale of civilization, and thus in some sense deservedly beyond memory or recall: 'I am ignorant, and I am careless, of the blind mythology of the Barbarians,' Gibbon wrote, in a spirit notably different from that great Italian thinker of the early Enlightenment, Vico. If Vico's concern had been to explain exactly how the savage mind had given birth to civilized thought, Gibbon by contrast habitually chose to underline the discontinuities between the primitive and the polite states. Although he admitted (acknowledging a long tradition of political analysis) that 'the most civilized nations of modern Europe issued from the woods of Germany,' he stressed that the consequence of the barbarian triumph was, over many centuries, massive retrogression. Its monuments were everywhere still visible in the ruins of civilization. Swine and buffalo were even now feeding in the forum in Rome; at Lambesa,

> a Roman city, once the seat of a legion, and the residence of forty thousand inhabitants, the Ionic temple of Aesculapius is encompassed with Moorish huts; and the cattle now graze in the midst of an amphitheatre, under the shade of Corinthian columns.

Nor had the destruction of civilization been confined to the effects of a single thunderclap; its calamities rumbled down the centuries, as Carthage's fate evidenced: 'in the beginning of the sixteenth century the second capital of the West was represented by a mosch, a college without students, twenty five or thirty shops, and the huts of five hundred peasants'. It was the cruel fate of Carthage to be ruined over and over again.

> Even that paltry village was swept away by the Spaniard whom Charles the Fifth had stationed in the fortress of the Goletta. The ruins of Carthage have perished; and the place might be unknown if some broken arches of an aqueduct did not guide the footsteps of the inquisitive traveller.

Civilization had fallen victim to the 'vicissitudes of fortune, which spares neither man nor the proudest of his works, which buries empires and cities in a common grave'. There has, of course, been much scholarly debate during the last century about whether Gibbon's picture of the severity of barbarization is correct. Historians such as Alfons Dopsch maintained that the Mediterranean commercial economy, urban life and education survived the irruption of the hordes. The Ostrogoths, Merovingians and Burgundians adapted themselves speedily and effectively, it was claimed, to the political order they had taken by force. As Peter Brown has noted, nineteenth-century historians schooled in the aspirations of Romantic nationalism were warm champions of the precocious capacities of the invaders. But Gibbon's vision of a fundamental hiatus of civilized life still carries a great deal of conviction.

A historian of the Roman Empire who stressed how civilization succumbed to barbarism might well have underwritten the age-old cyclical philosophy of history. In such a formulation of 'historical pessimism', one that left its mark on Volney and other contemporaries of Gibbon, barbarism would be seen as wiping out the gains of civilization, and mankind would have needed, Sisyphus-like, to start up afresh from the 'shipwreck of nations'.

The pessimism underlying this view of the destructive powers of time does indeed seem to inform Gibbon's interpretation of the fortunes of the Eastern empire. Because it lasted so much longer, Byzantium had the opportunity – which it eagerly seized – to sink more deeply into corruption than the West, a story ironically exemplified by the intellectual degradation (as Gibbon sees it) of Byzantium's own 'servile historians': civilization gets the historians it deserves. Gibbon's chapter 48, which gallops through the reigns of fifty increasingly decadent emperors in five progressively inglorious centuries, presents that vision of the East as a stagnant pool which was coming to dominate the Enlightenment mind in the West, anticipating the spirit of Tennyson's 'Better fifty years of Europe than a cycle of Cathay'. For Gibbon, what was Byzantine history but 'a tedious and uniform tale of weakness and misery'? –

> The subjects of the Byzantine empire, who assume and dishonour the names both of Greeks and Romans, present a dead uniformity of abject vices which are neither softened by the weakness of humanity nor animated by the vigour of memorable crimes.

Gibbon particularly indicted the Greek empire for having become a dead-end, bequeathing nothing to posterity:

> the fate of the Greek empire has been compared to that of the Rhine, which loses itself in the sands before its waters can mingle with the ocean. The scale of dominion is diminished to our view by the distance of time and place; nor is the loss of external splendour compensated by the nobler gifts of virtue and genius.

In Gibbon's fierce judgment, the real crime of the Eastern empire was that it had left behind no monuments of greatness and had contributed nothing to the future:

> In the revolution of ten centuries, not a single discovery was made to exalt the dignity or promote the happiness of mankind. Not a single idea had been added to the speculative systems of antiquity, and a succession of patient disciples became in their turn the dogmatic teachers of the next servile generation.

Gibbon saw in the fertility of its culture the index of the quality of a civilization. By that criterion, Byzantium was indeed unhealthy, even sterile:

> Not a single composition of history, philosophy or literature, has been saved from oblivion. ... The minds of the Greeks were bound in the fetters of a base and imperious superstition, which extends her dominion round the circle of profane science. Their understandings were bewildered in metaphysical controversy; in the belief of visions and miracles, they had lost all principles of moral evidence; and their taste was vitiated by the homilies of the monks, an absurd medley of declamation and scripture.

Moribund Byzantium had thus fallen into merited oblivion. In so far as Byzantine scholars did manage to preserve the memory of classical learning, their sole value to posterity had been to flee to the West in the teeth of the Turk, and help sow the seeds of the Renaissance in a new soil. Byzantium's longevity was purely negative; moreover, once it fell, it fell for good. In Gibbon's day, the Turk was still triumphant, and late eighteenth-century Ottoman civilization had little enough by way of culture, intellectual advances or system of government to commend it to cosmopolitan Europeans (its attractions lay elsewhere).

We may agree that Gibbon presents a travesty of Byzantium, though his distaste was one common in his day; as Horace Walpole told him, 'I am sorry you should have pitched on so disgusting a subject as the Constantinopolitan history.' All the same, in terms of his wider vision of the pulse of progress, he was not far off the mark. The civilization of Rome was finally destroyed in the Eastern empire and was not regenerated.

Not so in the West. For Gibbon's later volumes show Latin Christendom undergoing a notable ferment produced by the mixing of Rome and the Goths, leading to 'the gradual progress of Society from the lowest ebb of primitive barbarism to the full tide of modern civilization'. Admittedly, Gibbon of course found a lot of dead wood in the Middle Ages. Its intellectual life was stifled by the scholastic prattlings of 'the numerous vermin of mendicant friars, Franciscans, Dominicans, Augustins, Carmelites, who swarmed ... with habits and institutions variously ridiculous', and who thereby 'disgraced religion, learning and common sense'. Gibbon also declined to anticipate the later Romantic adulation of feudal chivalry, which he saw as mere barbarity viewed through rose-tinted spectacles: 'if heroism be confined to brutal and ferocious valour, Richard Plantagenet will stand high among the heroes of the age'.

Yet the tale of the West was also a story of recovery. Gradually, over the centuries, the rhythms of civilized life were restored, and indeed in some ways excelled the achievements of the empire; new forms of polity were evolved which, Gibbon insisted, actually provided bulwarks against a return of the excesses and evils of the imperial eagle. 'All that is human must retrograde if it do not advance,' insisted Gibbon, with an eye to the contrast between East and West; and in the new kingdoms of post-barbarian Europe Gibbon saw shoots of improvement as well as the withering of much that had borne fruit.

It is worth identifying four elements of Gibbon's account of the recovery of the West. First, the *Decline and Fall* is noteworthy for its relative absence of interest in papal history. Gibbon did not vilify the Papacy as that Beast which had so dominated Protestant demonology; nor did he credit it as the unwitting bearer of the torch of civilized government down the centuries of darkness. Gibbon mainly seems to be hinting at a 'Guelf' view, that the cunning of history ensured that the medieval Papacy aided freedom by resisting the imperial ambitions of the medieval Germanic Hohenstaufen monarchs. Even so, Gibbon never fully addressed the role of the medieval popes. *Prima facie* it seems a puzzling gap; but the omission is probably Gibbon's way of stressing that it would be wrong to regard the Papacy as the historical successor of Rome itself. In other words, as befits a man of the Enlightenment, Gibbon is more concerned with the relations between Antiquity and contemporary Europe than with a year-by-year chronicle of all the intervening events.

Second, Gibbon discerned, beneath the lines of that catalogue of destructive events which constituted the fall of civilization, one thread of continuity, indeed of improvement: the silent economic development of Europe over the centuries. Through trade, agriculture and manufactures, Europe had grown richer; once planted as victors within the Roman world, the barbarians had settled as agriculturalists, learning the use of money and gradually reweaving civilization's web. Agriculture formed a plateau of skills from which there was no backsliding, for the settling of the land created all kinds of permanent settlements, social, legal and political. Like the

Physiocrats, Gibbon pronounced the cultivation of land the basis of all true wealth and well-being.

Gibbon was supremely aware, of course, that traditional theologians and moralists for their part, and no less some of the more radical *philosophes*, believed that in the accumulation of wealth lay the seeds of moral and social corruption. But here he subscribed rather to Adam Smith and the other Scottish economists and moralists. A taste for 'luxury', and the development of an economic system which multiplied wants and the means of supplying them, were not necessarily evils. Far from it :

> such refinements, under the odious name of luxury, have been severely arraigned by the moralists of every age; and it might perhaps be more conducive to the virtue, as well as the happiness, of mankind, if all possessed the necessaries, and none the super-fluities, of life. But in the present imperfect condition of society, luxury, though it may proceed from vice or folly, seems to be the only means that can correct the unequal distributions of property. The diligent mechanic, and the skilful artist, who have obtained no share in the division of the earth, receive a voluntary tax from the possessors of land; and the latter are prompted, by a sense of interest, to improve those estates, with whose produce they may purchase additional pleasures.

Doubtless, luxury could be fatal to indolent peoples who abused their blessings; Gibbon notes of his own father, 'few minds have sufficient ressources to support the weight of idleness; and had he continued to walk in the path of mercantile industry, my father might have been happier and his son would be a richer man.' But the growth of wealth, far from being 'fatal' to 'an industrious people', would provide the surplus without undermining the discipline necessary for civilized life.

Unlike many 'country party' ideologues in Britain concerned to draw parallels with the present, Gibbon did not blame luxury as the prime cause of Rome's decline; in his view, true responsibility lay with the despotism stemming from the ambitions of emperors and the opportunities afforded by military might to seize power. Thus economic advance was relatively harmless and healthy. Furthermore, it was permanent, for the gains made by man the maker were never utterly lost; the arts of government might wax and wane, but the arts of the plough and the loom were cumulative.

Even late-medieval Europe was more advanced in quite tangible ways than the empire had been: Augustus 'had neither glass to his windows nor a shirt to his back'. Times had seen the useful arts and the civilizing process they fostered spread – albeit by accident and with costs – right around the globe: 'since the first discovery of the arts, war, commerce, and religious zeal have diffused, amongst the savages of the Old and New World, those inestimable gifts; they have been successively propagated; they can never be lost.' Gibbon's view blended a mild technological determinism with a vision of the benefits of industriousness which scotches the notion advanced by Fuglum and others that he was indifferent to contemporary economic developments:

> Each village, each family, each individual, must always possess both ability and inclination, to perpetuate the use of fire and of metals; the propagation and service of domestic animals; the methods of hunting and fishing; the rudiments of navigation; the imperfect cultivation of corn, or other nutritive grain; and the simple practice of the mechanic trades. Private genius and public industry may be extirpated, but these hardy plants survive the tempest, and strike an everlasting root into the most unfavourable soil. The splendid days of Augustus and Trajan were eclipsed by a cloud of ignorance, and the Barbarians subverted the laws and palaces of Rome. But the scythe, the invention or emblem of Saturn, still continued annually to mow the harvests of Italy; and the human feasts of the Laestrygons have never been renewed on the coast of Campania.

Thus prophets of doom and nostalgia-mongers such as Montesquieu, who had argued that the modern world was less densely populated than the Roman, were mistaken. Things had got better. 'This part of the world,' argued Gibbon, rapping Montesquieu over the knuckles for his statistically inept demographic pessimism, 'is incomparably more populated, more civilized, more wealthy, more enlightened, than it was then, and ... it is even much superior to what the Roman Empire was, Italy excepted.' Because of this inherent tendency of economies to develop, 'it may safely be presumed, that no people, unless the face of nature is changed' – and this was yet another of the extravagant fantasies of the frivolous Parisian intelligentsia – 'will relapse into their original barbarism.'

Gibbon's third main way of formulating the relations between past and present lay in pointing to the fact that – as with economic development – there was a certain cumulative element in the evolution of arts, literature and science – in short of civilized consciousness. Cultural achievement was, of course, in Gibbon's view the touchstone of a society's worth, and in this respect he saw Europe as pre-eminent. 'The Tartars have had a Jenghiz Khan, and the Goths an Alaric, but we turn our eyes from the bloodstained plains of Scythia to fix them with pleasure on Athens and Florence.' Within Gibbon's vision of the struggle between achievement and oblivion, letters proved more permanent than conquests – indeed it was only writing which kept the renown of conquests alive. 'The vain titles of the victories of Justinian are crumbled into dust: but the name of the legislator is inscribed on a fair and everlasting monument.' Even a novel might be destined for immortality: 'The romance of *Tom Jones*, that exquisite picture of human manners, will outlive the palace of the Escurial and the Imperial Eagle of the House of Austria.'

Imperial absolutism had been the euthanasia of the culture which it had inherited: Caesars checked activity and the levelling effects of arbitrary power necessarily quashed

> every generous attempt to exercise the powers, or enlarge the limits, of the human mind. The beauties of the poets and orators, instead of kindling a fire like their own, inspired only cold and servile imitations: or, if any ventured to deviate from those models, they deviated at the same time from good sense and propriety.

As a result, under the oppressive dominion of the empire, even under the Antonines, fear and sycophancy replaced the free spirit of inquiry and imagination: 'the name of Poet was almost forgotten, that of Orator was usurped by the sophists. A cloud of critics, of compilers, of commentators, darkened the face of learning.' The Dark Ages which followed did not wholly destroy the products of the mind. 'The mischances of time and accident have spared the classic works to which the suffrage of antiquity had adjudged the first place of genius and glory.' Finally, after many centuries, the 'revival of letters' recovered, rectified and reproduced the classics of Antiquity; and new works of the intellect appeared, thanks to the beneficial civic rivalries of the new states which had likewise emerged out of the barbarian darkness: 'on the revival of letters, the youthful vigour of the imagination after a long repose, national emulation, a new religion, new languages, and a new world, called forth the genius of Europe.' Whereas the empire had crushed and regimented the intellect, and 'the decline of genius was soon followed by the corruption of taste,' by contrast the new political order of Europe proved conducive to the flowering of the mind and the fostering of a new republic of letters. Gibbon had written during his student days in Lausanne: 'the cause of literature stands by freedom.' As with advances in the practical arts, we could look forward to continual improvements in the *beaux arts* and in *belles lettres*. Gibbon was amused to be able to make Scotland's transformation from rudeness to refinement serve as an example:

> if in the neighbourhood of the commercial and literary town of Glasgow, a race of cannibals has really existed, we may contemplate, in the period of the Scottish history, the opposite extremes of savage and civilized life. Such reflections tend to enlarge the circle of our ideas; and to encourage the pleasing hope that New Zealand may produce, in some future age, the Hume of the Southern Hemisphere.

Such hopes, Gibbon trusted, were not merely expressions of the endemic 'vanity of authors who presume the immortality of their name and writings'. Rather the modern world was founded on a fresh footing which afforded scope for that desire for mental independence ('the first wish of our heart ... the first blessing of our nature'), the new spirit of *sapere aude*, daring to know. The republic of letters of Gibbon's day ultimately derived from the Goths: 'They restored a manly spirit of freedom, and after the revolution of ten centuries, freedom became the happy parent of taste and science.' Gibbon thought his own were auspicious times for the progress of the inquiring mind. 'The merit of discovery has too often been stained with avarice, cruelty, and fanaticism; and the intercourse of nations has produced the communication of disease and prejudice,' he admitted. But now things were changing:

> A singular exception is due to the virtue of our own times and country. The five great voyages successively undertaken by the command of his present majesty were inspired by the pure and generous love of science and mankind. The same prince, adapting

his benefactions to the different stages of society, has founded a school in his capital, and has introduced into the islands of the South Sea, the vegetables and animals most useful to human life.

(This is a judgment not, as Graubard complains, filled with complacency, but ironically amused at the equation between the Royal Academy and potato-culture.) In this way the progressive life of the mind was closely associated for Gibbon with the fourth reason for taking an optimistic view of the overall course of civilization. For what secured modern freedom was the emergence out of the fragments of the old empire of a new European polity comprising a concert of smaller, independent states. To be precise:

> Europe is now divided into twelve powerful though unequal kingdoms, three respectable commonwealths, and a variety of smaller, though independent states; the chances of royal and ministerial talents are multiplied, at least, with the number of its rulers ... The abuses of tyranny are restrained by the mutual influence of fear and shame, republics have acquired order and stability; monarchies have imbibed the principles of freedom, or at least of moderation; and some sense of honour and justice is introduced into the most defective constitutions by the general manners of the times. In peace, the progress of knowledge and industry is accelerated by the emulation of so many active rivals; in war, the European forces are exercised by temperate and indecisive contests.

The benefits of such an interlocking fabric of rival states, and the balance of power thereby created, were great by contrast to the stifling, deadening uniformity of empire. Above all, the new order seemed to incorporate safeguards against the re-emergence of the spectre of world domination:

> The division of Europe into a number of independent states, connected, however, with each other, by the general resemblance of religion, language and manners, is productive of the most beneficial consequences to the liberty of mankind. A modern tyrant, who should find no resistance either in his own breast or in his people, would soon experience a gentle restraint from the example of his equals, the dread of present censure, the advice of his allies, and the apprehension of his enemies.

Of course, Gibbon was forced to admit, the equilibrium of a multitude of vying states was precarious, and 'the balance of power will continue to fluctuate, and the prosperity of our own or the neighbouring kingdoms may be alternately exalted or depressed.' Some element of instability was inevitable:

> but these partial events cannot essentially injure our general state of happiness, the system of arts, and laws, and manners, which so advantageously distinguish above the rest of mankind, the Europeans and their colonies.

For Gibbon, the superiority of Enlightenment Europe derived from having preserved, polished and propagated the legacy of Antiquity:

> the classics have much to teach ... The philosophers of Athens and Rome enjoyed the blessings and asserted the rights of civil and religious freedom. Their moral and political writings might have gradually unlocked the fetters of Eastern Despotism.

Under the conditions of the Byzantine regime, this could not happen. But the concept of freedom, the dignity of the citizen and the power of the independent mind, once translated to the West, worked a great ferment. Moreover, modern Europe's political system, depending as it did upon the existence of 'fixed and permanent societies, connected among themselves by laws and government, bound to their native soil by art and agriculture', and together securing an international balance of power, would ensure that this 'great republic' would not retrogress, that the stultifying effects of empire would not return. Thus, Gibbon contended, Europe had escaped from the endless cycles of despotic empires and barbarian cataclysms. Rome, the fourth and last world empire, had now ended; the reign of freedom had begun.

Not least, this was because Europe, advanced in arts and arms and wealth, was now 'secure from any future irruption of barbarians'. The nomadic hordes had now diminished, thanks to the rise of agriculture: in Asia 'the reign of independent barbarism' had been shrunk to 'the remnants of Calmucks and Uzbecks'. Above all, technological and military superiority – the gun – gave Europe a protection the Romans had never enjoyed, for nowadays 'before they can conquer, [barbarians] must cease to be barbarous'. In his assessment of the happy security of Europe, safe from barbarian invaders, Gibbon has been accused by recent historians of raising a straw man. But it is worth remembering that the notion of barbarian tribes swarming down from Asia still occasioned fashionable frissons to *philosophes* from Voltaire to Volney. The former dogmatized in his usual style about the return of barbarism:

> After raising itself for a time from one bog, it [the world] falls back into another; an age of barbarism follows an age of refinement. This barbarism in turn is dispersed, and then reappears; it is a continual alternation of day and night.

Gibbon, by contrast, was shrewd rather than visionary in his assessment of how the cycle of barbarism and civilization had actually been broken. Though much had been lost with Rome's decline and fall, to a man of the Enlightenment such as Gibbon, it seemed that Europe had in most respects now surpassed the Romans.

Gibbon espoused no dogma of natural, inevitable progress. Yet he was optimistic by the standard of his own day about the state of Europe, Britain in particular. Thanks to her unique constitution, Britain seemed to Gibbon, as to Hume, to have advantageously resolved the age-old dilemma of reconciling security with liberty: 'Britain is perhaps the only powerful and wealthy state which has ever possessed the inestimable secret of uniting the benefits of order with the blessings of freedom.' Gibbon has been accused of complacency, not least in failing to look beyond the fortunes of a relatively narrow stratum of society. There is an obvious truth in this; but one might simply say that he was a man who knew when to count his blessings:

> when I contemplate the common lot of mortality, I must acknowledge that I have drawn a high prize in the lottery of life. ... the double fortune of my birth in a free and enlightened country in an honourable and wealthy family is the lucky chance of an unit against millions.

Of course, Gibbon has often been castigated for offering trivial or myopic analyses of the benefits of the present. To some it seems almost comically anticlimactic that his main political message for his times was that Europe was safe from the terror of the Tartars. Yet the point is not so trivial. In the long term, a turning-point had been reached when civilization was no longer vulnerable to the superior numbers, bravery or ferocity of the 'barbarians'. Critics have also complained that Gibbon's reflections in the *Decline and Fall* neglected the barbarians within. Because of his own oligarchic complacency (we are told), he disdained to examine the condition of the people, to confront socio-economic unrest; failed first to spot the smoke-signals of the French Revolution on the horizon, and later to understand its significance as a world-historical phenomenon. For Fuglum, 'his fulminations against the French Revolution show an amazing lack of social insight.' Indeed, as he was the first to admit, the Revolution took him by surprise. On the other hand, it would have been highly peculiar if the critic of prophecy had set himself up as an oracle. He did not believe that history formed a springboard for predictions. At most, it might reveal affinities; but he found no exact historical parallels to events in France. As he confessed to his step-mother: 'What a strange wild world do we live in! You will allow me to be a tolerable historian, yet, on a fair review of ancient and modern times, I can find none that bear any affinity with the present.'

Gibbon's hostility to the Revolution was complete, and, from his stronghold as 'king of the Place' at Lausanne, he bandied around phrases such as 'wild visionaries', 'dangerous fanatics' and 'the new barbarians who labour to confound the order and happiness of society'. What perturbed this lover of freedom and enemy of absolute power was that through the Revolution a new tyranny would surely emerge, grounded not on the despotism of the individual monarch but on the will of the mob:

> The fanatic missionaries of sedition have scattered the seeds of discontent ... many individuals, and some communities, appear to be infected with the French disease, the wild theories of equal and boundless freedom.

It was a recipe for despotism.

Whatever we make of Gibbon's conversion to Burke, it would be silly to conclude that his fearful response to the French Revolution invalidates his account of the essential stability of the modern European polities. In his reactions to France, Gibbon was not a blind reactionary; the friend of the Neckers of course knew that France needed change. But he believed the true way forward was through the development of constitutional guarantees for liberty: 'Had the

French improved their glorious opportunity to erect a free constitutional Monarchy on their ruins of arbitrary power and the Bastille,
I should applaud their generous effort.' That opportunity had not
been seized, and the result was sure to prove disastrous. For 'a
people of slaves is suddenly become a nation of tyrants and canni-
bals.' Such a monstrous polity could not last, and history has vindi-
cated Gibbon's analysis. Jacobinism devoured itself, and Napoleon's
dream of uniting Europe under empire – the fifth monarchy? –
was overthrown by precisely that concert of European states in
which Gibbon had earlier shown such faith. Gibbon's account of
the superior political stability of modern Europe – one based upon
nation-states, advanced commercial society, the spread of property
and the balance of power – proved perceptive.

Gibbon had no metaphysical theory relating the course of history
to extra-historical goals. He did not accept, unlike some Renaissance
philosophers and Reformation theologians, that the human race
had degenerated ever deeper into decrepitude or depravity. Neither
did he believe, with Enlightenment optimists such as Priestley,
Godwin or Condorcet, that human nature was in a state of progressive transformation towards the millennium, religious or secular. Gibbon perceived no essential alteration in the nature of man,
that unstable amalgam of reason and passions, productive of dignity,
danger or disgrace. And yet human history was not simply an endless
series of cycles, wheeling around like the planets or a comet.

In one of the most evocative images ever conceived by a historian,
Gibbon used the appearance of a comet – Halley's comet – in the
reign of Justinian, to draw the contrast between the timeless cycles
of nature and the momentum of human history. That celestial body
had appeared to man seven times. It had never changed; but the
human society which had witnessed it had been successively transformed, responding to the appearance of this strange tailed star
in radically different ways. The first time, in 1767 BC (coeval with
Ogyges, the father of Grecian antiquity) it was taken as an un-
exampled portent. In 1193 BC, its appearance was inserted into the
myth of Electra; the third advent formed part of the Sibyl's omen;
at its fourth return, in 44 BC, it was interpreted by 'vulgar opinion'
as being the vehicle through which the soul of Julius Caesar would
be conveyed to the heavens.

Very gradually, such myths of deities, omens and heroes gave way, with the progress of the human mind, to more rational explanations. The fifth visitation, that during Justinian's reign, was still hailed as portending 'wars and calamities', but now at least some philosophical progress can be seen: 'the astronomers' explained that it was a 'planet of a longer period'. The sixth return, in AD 1106, was rationalized within the context of religious fanaticism: 'the Christians and the Mahometans might surmise, with equal reason, that it portended the destruction of the infidel.' Finally science and reason greeted the seventh return of the comet:

> The *seventh* phaenomenon of one thousand six hundred and eighty was presented to the eyes of an enlightened age. The philosophy of Bayle dispelled a prejudice which Milton's muse had so recently adorned, that the comet 'from its horrid hair shakes pestilence and war'. Its road in the heavens was observed with exquisite skill by Flamsteed and Cassini; and the mathematical science of Bernoulli, Newton, and Halley, investigated the laws of its revolutions.

Sweeping thus from pre-rational fable through fanaticism to a science which had finally revealed the principles of nature, Gibbon then offered an astonishing glimpse into the future:

> At the *eighth* period, in the year two thousand two hundred and fifty-five, their calculations may perhaps be verified by the astronomers of some future capital in the Siberian desert or American wilderness.

Human nature did not change, and men still burned with fierce passions and used reason as a sword to smite their enemies: Gibbon knew all that only too well from his Catholic past. Yet civilization nonetheless made gains, which were never totally lost. The age of the empires had given way to more durable polities which did not terrorize their subjects; economic improvement consolidated the fabric of settled, stable communities. Fanaticism had been curbed – civilization had disarmed the tempter. And reason, freedom and inquiry worked their ferment. Out of the terrifying clashes of uncontrolled forces emerged the impetus of improvement, and progress was the fortunate though unintended consequence of blind ambitions and monstrous dreams.

And improvement was real. Much as he admired his Romans, Gibbon knew that the sterile alternatives which faced a Cicero – discredited polytheism, bigoted zeal, hopeless Stoicism – had been transcended. Gibbon no longer had to be a Roman, but could bask as a man of the Enlightenment. That was a measure of betterment. We could indeed, as Gibbon claimed, 'acquiesce in the pleasing conclusion that every age of the world has increased, and still increases, the real wealth, the happiness, the knowledge, and perhaps the virtue of the human race'.

Conclusion: Making History

Gibbon provoked much wrath, above all among the Church Militant, furious at his use of 'grave and temperate irony even on subjects of ecclesiastical solemnity' ('who can refute a sneer?' demanded Archdeacon Paley, with some justice). But others were wrathful too. The rich and strange expatriate, William Beckford, wrote on the fly-leaf of his own copy of the *Decline and Fall*:

> The time is not far distant, Mr Gibbon, when your most ludicrous self-complacency, your numerous, and sometimes apparently wilful mistakes, your frequent distortion of historical Truth to provoke a gibe, or excite a sneer at every thing most sacred and venerable, your ignorance of the oriental languages, your limited and far from acutely critical knowledge of the Latin and the Greek, and in the midst of all the prurient and obscene gossip of your notes – your affected moral purity perking up every now and then from the corrupt mass like artificial roses shaken off in the dark by some Prostitute on a heap of manure, your heartless scepticism, your unclassical fondness for meretricious ornament, your tumid diction, your monotonous jingle of periods, will be still more exposed and scouted than they have been. Once fairly knocked off from your lofty, bedizened stilts, you will be reduced to your just level and true standard.

Few historians receive accolades like Beckford's, though a similar, albeit watered-down, vein of invective against the alleged defects of his character and his moral vision (if not his scholarship) runs through Victorian verdicts upon him. Gibbon's history evidently broke the mould and hurt many tender minds. So what was so scandalous about his history? And have Beckford's predictions about the imminent decline and fall of the author at last been realized? It is a fact, after all, that today's handbooks of historiography mention Gibbon only in passing; has he finally been reduced to his 'level and true standard'?

Surely the *Decline and Fall* personally offended its detractors, and seems out of line with the techniques and aspirations of today's professional historians, above all because Gibbon is such an unabashed presence – as historian, author and human being – in his 'corpulent' volumes. He quite explicitly placed himself within his own field of vision. At one point he devotes a footnote to expatiating upon the pleasures which writing the *Decline and Fall* has brought him. Solomon and the Caliph Abdalrahman, he tells us, counted their happy days but few. Yet:

> Their expectations are commonly immoderate, their estimates are seldom impartial. If I may speak of myself (the only person for whom I can speak with certainty), my happy hours have far exceeded, and far exceed, the scanty number – fourteen days – of the caliph of Spain; and I shall not scruple to add that many of them are due to the pleasing labour of the present composition.

And he makes himself part of his story in more oblique ways too. Rather like an artist slipping his self-portrait into a crowd scene, Gibbon offers towards the close of his work a vignette of the scholar Barlaam, which, with its ineffable blend of vanity and self-mocking irony, demands to be taken as a portrait of the artist as a man of learning:

> He is described by Petrarch and Boccace, as a man of diminutive stature, though truly great in the measure of learning and genius; of a piercing discernment, though of a slow and painful elocution. For many ages (as they affirm) Greece had not produced his equal in the knowledge of history, grammar, and philosophy; and his merit was celebrated in the attestations of the princes and doctors of Constantinople.

It is a nice stroke of self-mockery that Gibbon takes his final bow dressed up as a Byzantine. In any case, no reader of the *Decline and Fall* will forget the implied presence of Gibbon 'musing amidst the ruins of the Capitol, while the barefooted friars were singing vespers in the Temple of Jupiter', conceiving 'the idea of writing the decline and fall of the city', or the haunting passage in the *Memoirs* in which Gibbon, alone in Lausanne, recorded his 'deliverance':

It was on the day, or rather night of the 27th of June 1787, between the hours of eleven and twelve, that I wrote the last lines of the last page, in a summer-house in my garden. After laying down my pen, I took several turns in a *berceau*, or covered walk of acacias, which commands a prospect of the country, the lake and the mountains. The air was temperate, the sky was serene, the silver orb of the moon was reflected from the waters, and all nature was silent. I will not dissemble the first emotion of joy on the recovery of my freedom and, perhaps, the establishment of my fame. But my pride was soon humbled, and a sober melancholy was spread over my mind, by the idea that I had taken an everlasting leave of an old agreable companion, and that whatever might be the future fate of my *History*, the life of the historian must be short and precarious.

Gibbon thus makes himself part of the story. Machiavelli tells us that to get himself into the mood for writing his *Il Principe*, he would put on 'the robes of court and palace, and in this graver dress I enter the antique courts of the ancients'. In a rather similar act of identification, Gibbon came to think of himself as a patrician, bewailing the triumph of barbarism and religion all around him – or even as a general: one morning at half past seven, he writes, 'as I was destroying an army of Barbarians', cousin Eliot dropped by to offer a parliamentary seat.

Gibbon thus associates himself with his book ('an old and agreable companion'). Yet his self-inclusion also serves as a vitally important distancing device. By placing himself within the story, by drawing attention to himself – as narrator, as judge, as the beaming author of all those sly and lubricious footnotes – he is continually reminding the reader of writing in the *present*, as well as absorbing him in the *past*. The *Decline and Fall* has been said to draw heavily upon the models of the epic and tragic theatre. This is so. But Gibbon's masterly balancing act, whereby the author is at once creator and a character of his own creation ('the historian of the Roman Empire', a persona adopted from the third volume onwards), is indebted not primarily to such classical genres, but to the novel; above all, to that work for which Gibbon so confidently predicted immortality, *Tom Jones*, but also, one suspects, to *Tristram Shandy*.

For the inexhaustible fascination of the *Decline and Fall*, and its true value to our understanding of the historian's craft, lies in

Gibbon's exceptionally acute self-awareness of the persona of the historian. It is a consciousness expressed not in methodological pro-legomena, nor even in *parodic* methodological prolegomena – Gibbon's sense of humour wasn't Swiftian – but through the exquisite artistry of his imagination in creating the historian, the history and no less his readership.

Long before Croce and Collingwood, Gibbon knew that all history is contemporary history, and furthermore enjoyed the little conceits which followed upon the conflation of the past and the present. He knew that his references to the revolt against Rome by the *Armorican* provinces (i.e. north-west Gaul: roughly Brittany and Normandy) were bound to be read as a commentary on the revolt of the *American* colonies. (He might have chuckled at disputes among historians today about whether that was what he intended.) He must have fluttered with pride when passages from the *Decline and Fall* were actually read out in Parliament as lessons in civil prudence – thereby incidentally completing the circle, for Gibbon had won the civil prudence essential for the historian through sitting as a 'senator' on the back benches. 'The present is a fleeting moment,' he wrote, 'the past is no more'; the task of the historian was to recapture this lost past within the fleeting present.

Gibbon always regarded history as 'a liberal and useful study', and it has often been claimed by scholars, Americans in particular, that Gibbon was instructing the British Empire when he wrote of the Roman: 'Gibbon was searching for literal analogies,' writes Stephen Graubard. Yet he did not crassly believe that history repeated itself; Gibbon knew that, though passions are timeless, circumstances are never the same. The politics of Rome might illuminate the politics of England, but it would be a travesty to imply that Gibbon crudely conflated the Roman Empire and the British, or more generally that he saw history as a kind of eternal return, and history books like *Old Moore's Almanac*. History never repeated itself, for the power of consciousness, not least consciousness of the past, generated cumulative change, or distance at least.

Gibbon sought above all to be a philosophical historian. 'Philo-sophical' method meant many things for him, including all the laud-able skills involved in training one's mind on the evidence: con-textualizing, interpreting, evaluating, reflecting and so forth. It also implied a degree of detachment. This was not in aid of some Stoic goal of transcending a past which was little more than a 'register of the crimes, follies and misfortunes of mankind': Gibbon was never slow to puncture the pretensions of those who aspired to transcend the limits of human nature.

Nor was it Gibbon's aim to achieve dispassionate neutrality, the kind of scientific objectivity which has so often been commended to scholars over the last century and a half as the epistemological acme of the historian (in Bury's words, 'history is a science, no less and no more'). It was the goal of 'scientific history', as conceived by the great Ranke, to turn the historian into an invisible man, and let the facts, the sources, speak for themselves. It was a noble aim. But Gibbon would, rightly, have seen it as a terrible delusion: how could an author ever write himself out of his work?

Rather, philosophical detachment was necessary, Gibbon believed, because it enabled the historian to confront, interpret and not least express the radical contingency, even the mysteries, of the historical stage. The philosophical temper would equip a scholar for understanding that history was not an answer-book, nor even a compendium of sociological 'laws'. It was an endless succession of engagements with a past in which the *dramatis personae* themselves were never able fully to fathom, control and command events. Historians might be wiser, but only because they could recognize the plight of their protagonists, not because they could know the final truth. Thinking history was an unfinished series of honourable engagements, between the passion to understand and the prompting of reason that we must rest content with a modest scepticism. The philosophical historian must pick his path between the presumption of omniscience and the despair of ignorance.

Merely to grasp this was not enough. The historian's craft required him to find modes of expressing it. Gibbon understood like few others that it was indeed the historian who made history. If ever a historian paid due tribute to the role of imagination ('always active to enlarge the narrow circle in which nature has confined us') in the making of historical truth, it was Gibbon, whose notion of the act of historical creation properly emphasized 'the sportive play of fancy and learning'.

But imagination was not enough. An art of expression was needed to do full justice to the radical uncertainty and multiplying meanings which the philosophical historian perceived. Gibbon appreciated the usefulness of literary devices in giving form to the facts, pointing parallels, suggesting allusions, indicating the rhythms of affairs, and showing how both actor and event were contained within the larger whole. To achieve this Gibbon above all perfected the use of irony; by using language to open up the multiplication of meanings, he wanted the reader to be unsettled, brought face to face with that register of responses present in the historian's mind. Irony disabuses the reader of the illusion that history contains solutions.

By placing himself both within his history and as a lofty spectator beyond it, Gibbon established that double vision which is the hall-mark of the master historian. He would evoke, but he would also judge; and be fully aware that the historian must perform many parts on the stage he has created. He knew that pure history liberated from subjectivity was a chimera, and attempts to find it led at best to dullness and at worst to delusion.

Gibbon advocated impartiality. By this we must not understand any speciously scientist notion of neutrality, but rather an endea-vour, through cultivating self-awareness, to rise above partisanship, towards the pinnacle of the 'philosophical observer'. There was no 'solution' to the problem of bias. Gibbon coped with bias by revealing it; by exposing the prejudices of his sources, and present-ing, rather than suppressing, the personality of 'the historian of the Roman Empire'. Impartiality arose not out of a fetishism of facts but from the operations of the mind, from analysis, imagina-tion, wit and a capacity to hold judgment in suspense. Gibbon's 'great work' reads like a chorus of voices. The contemporaries speak; Gibbon's sources comment on them; Gibbon adds his glosses, often scolding away in the footnotes; and the reader is invited to listen and participate in the intellectual symposium. Not the least element of Gibbon's craft of history is to include, indeed to implicate, the reader within his theatre of the world. The reader seems to be a missing person in so much history-writing today, presumably because academic historians have largely lost all sense of writing for a public. Gibbon, by contrast, enters into a compact with his audience, often directly addressing them: 'Read and feel the XXIIId book of the Iliad,' he tells them at one point.

In some ways, Gibbon's priorities for the historian were very much those of his age. The literary world of the eighteenth century, with its unstable and rapidly changing position for the writer, encouraged the projection of such highly self-conscious, indeed pre-carious, personae for the author. Yet a comparison of the *Decline and Fall* with other histories then rolling off the presses reveals that Gibbon was also unique in Britain, a philosophical historian in a manner matched only, if at all, by Hume.

That achievement of philosophical detachment, that sense of history personal no less than history epic, as infinitely complex and profoundly inscrutable, surely arose from the bewildering nature of a life so often but so falsely called eventless. If he was not *sui generis*, Gibbon was at least an oddity: traversing between England and Switzerland; now a 'patriot', now a 'citizen of the world'; now an Anglican, now a Catholic, now 'good Protestant', and then sliding 'from superstition to scepticism'. Not many historians had to keep reinterpreting the world from so many different vantage points. Few ended up with the intellectual freedom of the bachelor scholar with his comfortable competence; but few had to battle so stressfully to gain such a position – and to jettison so much in the process. Gibbon emerged as a scholar gentleman; how many literary gentlemen had such burning scholarly ambitions? It is not surprising that Gibbon's work has a scale and universality unmatched by his historical successors.

I have been suggesting why Gibbon's idea of history was different from the academic formulations of later generations. Gibbon's preoccupations have little to do with the debates about causality, scientific laws, historicism and the philosophical status of explanation in history, which were argued by nineteenth-century idealists, positivists and evolutionists, and which have constituted hard-core 'philosophy of history' ever since. Rather he was fascinated by history as the creation of the historian's mind playing upon the mind of the reader and passionately concerned about its capacity to enlighten, entertain, interest and instruct.

History moved on. In the next generations, Mitford and Grote wrote histories of Greece which in learning at least approximated what Gibbon had done for Rome. Romanticism transformed historical sympathies. Scott brought the Middle Ages into high fashion, and new schools of nationalist historians narrowed the astonishing catholicity characteristic of Gibbon. With the assimilation of Germanic scholarship, superior techniques opened historians' eyes to fresh sources and ways to interpret them. Yet no later British historian produced an account of the course of history from Antiquity through the Middle Ages into modernity to match the *Decline and Fall*. No one superseded Gibbon as 'the historian of the Roman Empire'.

Bibliographical Essay

This essay serves two functions. It is intended to indicate the main works of scholarship upon which the argument and analysis in this book are based. It is also meant to be a guide for further reading. I have arranged it so that, as far as possible, it corresponds to the topics covered in the respective chapters.

First, however, a word is necessary about Gibbon's own writings. There are many abridged editions of the *Decline and Fall* available. Easily the best complete text remains the one edited by J. B. Bury in seven volumes (London, Methuen, 1909–14). It contains an informative introduction, and excellent notes which correct Gibbon in the light of nineteenth-century scholarship. Most of Gibbon's other scholarly writings are to be found in his *Miscellaneous Works*, edited by Lord Sheffield, 5 volumes (London, John Murray, 1814); more accessible is P. B. Craddock, *The English Essays of Edward Gibbon* (Oxford, Clarendon Press, 1972). This volume includes some hitherto unprinted writings from manuscripts in the British Library. The most convenient scholarly edition of Gibbon's autobiography, splicing all six variants into a single narrative, is G. A. Bonnard (ed.), *Edward Gibbon, Memoirs of My Life* (London, Nelson, 1966). Gibbon's journals have also been edited: David M. Low (ed.), *Gibbon's Journal to January 28th, 1763: My Journal, I, II, & III and Ephemerides* (London, Chatto & Windus, 1929), and G. A. Bonnard (ed.), *Gibbon's Journey from Geneva to Rome: His Journal from 20th April to 2nd October 1764* (London, Nelson, 1961). The best edition of his letters is J. E. Norton (ed.), *The Letters of Edward Gibbon*, 3 volumes (London, Cassell, 1956), though R. E. Prothero (ed.), *Private Letters of Edward Gibbon, 1753–1794*, 2 volumes (London, John Murray, 1896) is still worth consulting, since it includes letters sent to Gibbon as well as ones written by him. Geoffrey Keynes, *The Library of Edward Gibbon: A Catalogue of His Books* (London, The Bibliographical Society, 1950) offers an excellent account of the books Gibbon owned, and thus a survey of his reading habits.

There is no good complete up-to-date survey of Gibbon scholarship. A good bibliography, running up to the mid-1970s, is to be found in Michel Baridon, *Edward Gibbon et le mythe de Rome: Histoire et idéologie au siècle des lumières*, 2 volumes (Paris, Editions Honoré

Champion, 1977) – a monumental work which has suffered neglect, one suspects because it has never been translated.

Introduction

Most of the works on Gibbon casually referred to in the Introduction are listed in other sections of this Essay. The severe verdict of Sir Geoffrey Elton is taken from *The Practice of History* (Sydney, Sydney University Press, 1967). Most other scholars who have written historical primers have a few words to say about Gibbon. He does not loom large, however, in standard texts on the philosophy and methodology of history, precisely because the concerns of recent methodology (focusing upon scientific method and epistemological issues) are so different from Gibbon's own preoccupations (mainly with authorship).

The zealous Freudian interpretation I cite comes from Willis R. Buck, 'Reading Autobiography', *Genre*, 13 (1980), pp. 477–98. Martine Watson Brownley's negative comments are in her 'Gibbon's Artistic and Historical Scope in the *Decline and Fall*', *Journal of the History of Ideas*, 42 (1981), pp. 629–42.

Chapter 1: The Uses of History in Georgian England

Little has been written about the history of history-writing in eighteenth-century England. The earlier period, when historical scholarship emerged out of Renaissance humanism, has however been well covered in J. R. Hale, *The Evolution of British Historiography* (Cleveland, Meridian Books, 1964); Herschel Baker, *The Race Against Time* (Toronto, University of Toronto Press, 1967); A. Ferguson, *Clio Unbound* (London, Duke University Press, 1979); F. Smith-Fussner, *The Historical Revolution: English Historical Writing and Thought, 1580–1640* (London, Routledge & Kegan Paul, 1962); Levi Fox (ed.), *English Historical Scholarship in the Sixteenth and Seventeenth Centuries* (Oxford, Oxford University Press, 1956) and, most recently, Joseph Levine, *Humanism and History: Origins of Modern English Historiography* (Ithaca, Cornell University Press, 1987), which includes illuminating essays on the Ancients *versus* Moderns debate and on Gibbon's debt to humanism. Paul Fussell, *The Rhetorical World of Augustan Humanism* (Oxford, Oxford University Press, 1965); J. W. Johnson, *The Formation of English Neo-Classical Thought* (Princeton, Princeton University Press, 1967); and John Barrell, *The Political Theory of Painting from Reynolds to Hazlitt: 'The Body of the Public'* (New Haven, Yale University Press, 1986) offer excellent insights on the continuing attractions of humanism for the eighteenth-century mind.

There is some fine scholarship upon the ideology of civic humanism and its appropriation of a usable past. Particularly valuable are J. G. A. Pocock, 'Between Machiavelli and Hume: Gibbon as Civic Humanist and Philosophical Historian', in G. W. Bowersock, John Clive and Stephen R. Graubard (eds.), *Edward Gibbon and the Decline and Fall of the Roman Empire* (Cambridge, Massachusetts, Harvard University Press, 1977), pp. 103–20; and his *The Machiavellian Moment: Florentine Political Thought and the Atlantic Tradition* (Princeton, Princeton University Press, 1975); Isaac Kramnick, 'Augustan Politics and English Historiography', *History and Theory*, 6 (1967), pp. 33–56; and Frank Turner, 'British Politics and the Demise of the Roman Republic: 1700–1939', *The Historical Journal*, 3 (1986), pp. 577–99. For Gibbon as a politician see S. Lutnick, 'Edward Gibbon and the Decline and Fall of the First English Empire: The Historian as Politician', *Studies in Burke and His Time*, 10 (1967–68), pp. 1097–1113.

The idea of a sacred history of mankind can be approached generally in Leslie Stephen, *History of English Thought in the Eighteenth Century*, 2 volumes (reprinted, London, Harbinger, 1962); G. R. Cragg, *Reason and Authority in the Eighteenth Century* (Cambridge, Cambridge University Press, 1964); and C. A. Patrides, *The Phoenix and the Ladder: The Rise and Decline of the Christian View of History* (Berkeley, University of California Press, 1964). A powerful account of one specific instance is to be found in Frank E. Manuel, *Isaac Newton, Historian* (Cambridge, Cambridge University Press, 1963). The broader debate, focusing on the rivalry between Christian and naturalistic histories of creation, is surveyed in Paolo Rossi, *The Dark Abyss of Time*, trans. L. G. Cochrane (Chicago, University of Chicago Press, 1984), and in F. E. Manuel, *The Changing of the Gods* (Hanover, Brown University Press, 1983), and his *The Eighteenth Century Confronts the Gods* (Cambridge, Massachusetts, Harvard University Press, 1959). See also Paul Hazard, *La Crise de la conscience Européene* (Paris, Boivin, 1935), and also his *La Pensée Européene au XVIIIe siècle* (Paris, Boivin, 1946).

The most helpful introduction to writers, readers and the literary market place are A. S. Collins, *Authorship in the Days of Johnson* (London, R. Holden, 1927) and J. W. Saunders, *The Profession of English Letters* (London, Routledge & Kegan Paul, 1964). The plight of history at the English universities is covered in L. S. Sutherland and L. G. Mitchell (eds.), *The History of the University of Oxford* (general editor, T. H. Aston), vol. 5, *The Eighteenth Century* (Oxford, Clarendon Press, 1986) and J. Gascoigne, *Science and Religion in Georgian Cambridge* (Cambridge, Cambridge University Press, 1988).

Lastly, there are some helpful accounts of eighteenth-century British historiography. J. B. Black, *The Art of History* (London, Methuen, 1926) has useful essays on Hume, Robertson and Gibbon; P. Thomas Peardon, *The Transition in English Historical Writing, 1760–1830* (New York, Columbia University Press, 1933) offers a workmanlike survey of the transition from rationalist to Romantic history; Forbes's introduction to Hume contains brilliant analysis: David Hume, *The History of Great Britain*, edited by Duncan Forbes (Harmondsworth, Penguin, 1970); and see also R. N. Stromberg, 'History in the Eighteenth Century', *Journal of the History of Ideas*, 12 (1951), pp. 295–404; M. A. Thomson, *Some Developments in English Historiography during the Eighteenth Century* (London, H. K. Lewis, 1957); Hugh Trevor-Roper, 'The Historical Philosophy of the Enlightenment', *Studies on Voltaire and the Eighteenth Century*, 27 (1963), pp. 1667–87; G. H. Nadel, 'Philosophy of History before Historicism', in G. H. Nadel (ed.), *Studies in the Philosophy of History* (New York, Harper & Row, 1965), pp. 49–73; Herbert Davis, 'The Augustan Conception of History', in J. A. Mazzeo (ed.), *Reason and the Imagination* (New York, Columbia University Press, 1962). D. C. Douglas, *English Scholars (1660–1730)* (London, Jonathan Cape, 1943) illuminates the antiquarian temper, and Philippa Levine, *The Amateur and the Professional: Antiquarians, Historians and Archaeologists in Victorian England, 1838–1886* (Cambridge, Cambridge University Press, 1986) is helpful on the later professionalization of history. Gibbon's attempt to get British sources printed is examined in Hugh Trevor-Roper, 'The Other Gibbon', *American Scholar*, 46 (1976–7), pp. 94–103. Finally there is an admirably stimulating comprehensive survey of the rise of history in Britain in John Kenyon's *The History Men* (London, Weidenfeld & Nicolson, 1983), though Gibbon, so worried about losing his name, would have been disconcerted to find himself making his first appearance there as 'Edmund'.

Chapter 2: The Making of the Historian

There are some excellent biographies. General surveys in the traditional mould include D. M. Low, *Edward Gibbon* (London, Chatto & Windus, 1937); G. M. Young, *Gibbon* (London, Rupert Hart-Davis, 1948); and rather more recently, R. N. Parkinson, *Edward Gibbon* (New York, Twayne Publishers, 1973); Edward Joyce, *Edward Gibbon* (London, Longmans, Green & Co., 1953); and Joseph Ward Swain, *Edward Gibbon the Historian* (London, Macmillan, 1966). Sir Gavin de Beer, *Gibbon and His World* (London, Thames & Hudson, 1968) offers a well-illustrated coffee-table life and times. Two recent works have concentrated on the development of Gibbon's mind from adolescence to the maturity of the 'historian of the Roman Empire': Patricia B. Craddock, *Young Edward Gibbon: Gentleman of Letters* (Baltimore, The Johns Hopkins University Press, 1982) and David P. Jordan, *Gibbon and His Roman Empire* (Chicago, University of Illinois Press, 1971). Craddock makes particularly fine use of manuscript material. Though it suffers from wild speculation and unbalanced judgments, Lionel Gossman, *The Empire Unpossess'd: An Essay on Gibbon's Decline and Fall* (Cambridge, Cambridge University Press, 1981) is sensitive to how the young Gibbon tried to make sense, through the past, of his own present and future.

Gibbon's *Memoirs*, and the problems of their interpretation, have also attracted some fine scholarship. Amongst the more sensitive accounts are Patricia Meyer Spacks, *Imagining A Self: Autobiography and Novel in Eighteenth-Century England* (Cambridge, Massachusetts, Harvard University Press, 1976), pp. 92–126; Martin Price, 'The Inquisition of Truth: Memory and Freedom in Gibbon's Memoirs', *Philological Quarterly*, 54 (1975), pp. 391–407; Robert H. Bell, 'Gibbon: The Philosophic Historian as Autobiographer', *Michigan Academician*, 13 (1981), pp. 349–64; and Robert Folkenflik, 'Child and Adult: Historical Perspective in Gibbon's Memoirs', *Studies in Burke and His Time*, 15 (1973–74), pp. 31–43. I have not had space to explore the very real problems of interpreting Gibbon's life through a multiplicity of autobiographies. These problems are intelligently surveyed by Spacks.

The significance of the *Essai* and other early writings for Gibbon's intellectual development is well brought out in G. Giarrizzo, 'Toward the *Decline and Fall*: Gibbon's Other Historical Interests', in G. W. Bowersock, John Clive and Stephen R. Graubard (eds.), *Edward Gibbon and the Decline and Fall of the Roman Empire* (Cambridge, Massachusetts, Harvard University Press, 1977), pp. 233–46; Jean Starobinski, 'From the Decline of Erudition to the Decline of Nations: Gibbon's Response to French Thought', in G. W. Bowersock, John Clive and Stephen R. Graubard (eds.), *Edward Gibbon and the Decline and Fall of the Roman Empire* (Cambridge, Massachusetts, Harvard University Press, 1977), pp. 139–58; Frank E. Manuel, 'Edward Gibbon: Historien-Philosophe', in G. W. Bowersock, John Clive and Stephen R. Graubard (eds.), *Edward Gibbon and the Decline and Fall of the Roman Empire* (Cambridge, Massachusetts, Harvard University Press, 1977), pp. 167–81; Gibbon's debt to Locke is addressed in Curt Hartog, 'Gibbon and Locke', *Genre*, 61 (1982), pp. 415–29. The Swiss history is examined in H. S. Offler, 'Gibbon and the Making of His Swiss History', *Durham University Journal*, 41 (1949), pp. 64–74. P. R. Ghosh, 'Gibbon's Dark Ages: Some Remarks on the Genesis of the *Decline and Fall'*, *The Journal of Roman Studies*, 73 (1983), pp. 1–23, is the most valuable attempt yet to date Gibbon's unpublished writings.

Chapter 3 : *The Decline and Fall*

There are few detailed analyses of the *Decline and Fall* itself. Quality makes up for quantity, however, and both G. Giarrizzo, *Edward Gibbon e la Cultura Europea del Settecento* (Naples, Istituto Italiano par gli Studi Storici, 1954) and Michel Baridon, *Edward Gibbon et le mythe de Rome: Histoire et idéologie au siècle des lumières* (Paris, Editions Honoré Champion, 1977) can be warmly recommended. Much less incisive is Per Fuglum, *Edward Gibbon: His View of Life and Conception of History* (Oxford, Basil Blackwell, 1953). The epic structure of the work is brought out by Leo Braudy, *Narrative Form in History and Fiction* (Princeton, Princeton University Press, 1970), and E. M. W. Tillyard, *The English Epic and its Background* (Oxford, Oxford University Press, 1966). Gibbon's own affinities with the classical past are discussed in C. N. Cochrane, 'The Mind of Edward Gibbon', *University of Toronto Quarterly*, 12 (1942), pp. 1–17, and 13 (1943), pp. 146–66; Carl L. Becker, *The Heavenly City of the Eighteenth-Century Philosophers* (New Haven, Yale University Press, 1962) pictures Gibbon as an ancient Roman. The parallels and contrasts with Voltaire emerge well from J. H. Brumfitt, *Voltaire Historian* (Oxford, Oxford University Press, 1958); and Gibbon's relations to Montesquieu can be gauged from Robert Shackleton, 'The Impact of French Literature on Gibbon', in G. W. Bowersock, John Clive and Stephen R. Graubard (eds.), *Edward Gibbon and the Decline and Fall of the Roman Empire* (Cambridge, Massachusetts, Harvard University Press, 1977), pp. 207–18. A sympathetic interpretation of Gibbon as a philosophical historian is offered in J. H. Plumb, 'Gibbon and History', *History Today*, 19 (1969), pp. 737–43.

The quality of Gibbon's historical scholarship is assessed in Arnaldo Momigliano, 'Gibbon's Contribution to Historical Method', *Studies in Historiography* (London, Garland, 1966), pp. 40–55, and I. W. J. Machin, 'Gibbon's Debt to Contemporary Scholarship', *Review of English Studies*, 15 (1939), pp. 84–88.

The most penetrating insights into Gibbon's style are contained in Peter Gay, *Style in History* (New York, Basic Books, 1974). A lengthier and more formal survey is Harold L. Bond, *The Literary Art of Edward Gibbon* (Oxford, Clarendon Press, 1960). John Clive, 'Gibbon's Humor', in G. W. Bowersock, John Clive and Stephen R. Graubard (eds.), *Edward Gibbon and the Decline and Fall of the Roman Empire* (Cambridge, Massachusetts, Harvard University Press, 1977), pp. 183–

92 is a witty essay on Gibbon's wit. Martine Watson Brownley, 'The Theatrical World of the "Decline and Fall"', *Papers on Language and Literature*, 15 (1979), pp. 263–77, probes Gibbon's vision of the artificial: see also her 'Appearance and Reality in Gibbon's History', *Journal of the History of Ideas*, 38 (1977), pp. 651–66.

Chapter 4 : Power

Gibbon's views on politics and power have recently come under scrutiny. His commitment to liberty, deep distrust of imperial power, and refusal to romanticize barbarism are exceptionally well brought out in J. G. A. Pocock, 'Gibbon's *Decline and Fall* and the World View of the Late Enlightenment', *Eighteenth Century Studies*, 10 (1977), 287–303, and his 'Between Machiavelli and Hume: Gibbon as Civic Humanist and Philosophical Historian', in G. W. Bowersock, John Clive and Stephen R. Graubard (eds.), *Edward Gibbon and the Decline and Fall of the Roman Empire* (Cambridge, Massachusetts, Harvard University Press, 1977), pp. 103–20. On these issues, as on so many others, J. W. Burrow, *Gibbon* (Oxford, Oxford University Press, 1985) positively glows with illumination. Also useful are G. J. Gruman, 'Balance and Excess as Gibbon's Explanation of the Decline and Fall', *History and Theory*, 1 (1960), pp. 75–85; A. Lentin, 'Edward Gibbon and "The Golden Age of the Antonines"', *History Today* (July 1981), pp. 33–37; and, on the barbarians, Jeffrey B. Russell, 'Celt and Teuton', in Lynn White, Jr (ed.), *The Transformation of the Roman World* (Los Angeles, University of California Press, 1966), pp. 232–65. Georgian attitudes towards barbarians are surveyed in Samuel Kliger, *The Goths in England* (Cambridge, Massachusetts, Harvard University Press, 1952).

Gibbon's view of Byzantium is critically evaluated in Deno J. Geanakoplos, 'Edward Gibbon and Byzantine Ecclesiastical History', *Church History*, 35 (1966), pp. 170–85; Steven Runciman, 'Gibbon and Byzantium', in G. W. Bowersock, John Clive and Stephen R. Graubard (eds.), *Edward Gibbon and the Decline and Fall of the Roman Empire* (Cambridge, Massachusetts, Harvard University Press, 1977), pp. 53–60; and Speros Vryonis, 'Hellas Resurgent', in Lynn White, Jr (ed.), *The Transformation of the Roman World* (Los Angeles, University of California Press, 1966), pp. 92–118.

Chapter 5 : Religion

Gibbon's religious odyssey can be traced in the standard biographies. The best interpretation of his views on the psychology of religion is offered by J. G. A. Pocock, 'Superstition and Enthusiasm in Gibbon's History of Religion', *Eighteenth Century Life*, 8 (1982), pp. 83–94.

Clerical antagonism to him is fully documented in Shelby T. McCloy, *Gibbon's Antagonism to Christianity* (London, Williams & Norgate, 1933). Gibbon's theological erudition is examined by Owen Chadwick, 'Gibbon and the Church Historians', in G. W. Bowersock, John Clive and Stephen R. Graubard (eds.), *Edward Gibbon and the Decline and Fall of the Roman Empire* (Cambridge, Massachusetts, Harvard University Press, 1977), pp. 219–32. For Islam see G. E. Grunebaum, 'Islam: The Problem of Changing Perspective', in Lynn White, Jr (ed.), *The Transformation of the Roman World* (Los Angeles, University of California Press, 1966), pp. 147–78; and Bernard Lewis, 'Gibbon on Muhammad', in G. W. Bowersock, John Clive and Stephen R. Graubard (eds.), *Edward Gibbon and the Decline and Fall of the Roman Empire* (Cambridge, Massachusetts, Harvard University Press, 1977), pp. 61–74.

Peter Gay, *The Enlightenment: An Interpretation*, 2 volumes (London, Weidenfeld & Nicolson, 1966), vol. 1, places Gibbon in the context of 'the rise of modern paganism'. Christopher Dawson, 'Edward Gibbon', *Proceedings of the British Academy*, 20 (1934), pp. 159–80 offers an intelligent critique of Gibbon's supposed misunderstanding of Christianity. Gibbon's 'infidelity' has however been questioned in Paul Turnbull, 'The Supposed Infidelity of Edward Gibbon', *The Historical Journal*, 5 (1982), pp. 23–51.

Chapter 6: Civilization, Barbarism and Progress

The interplay of Rome, civilization and barbarism in Gibbon's thought emerges from the works cited in the bibliography to the two previous chapters. See also François Furet, 'Civilization and Barbarism in Gibbon's History', in G. W. Bowersock, John Clive and Stephen R. Graubard (eds.), *Edward Gibbon and the Decline and Fall of the Roman Empire* (Cambridge, Massachusetts, Harvard University Press, 1977), pp. 159–66. For Gibbon's analysis of barbarism see J. G. A. Pocock, 'Gibbon and the Shepherds: The Stages of Society in the *Decline and Fall*', *History of European Ideas*, 2 (1981), pp. 193–200. Particularly valuable is Peter Brown, 'Gibbon's Views on Culture and Society in the Fifth and Sixth Centuries', in G. W. Bowersock, John Clive and Stephen R. Graubard (eds.), *Edward Gibbon and the Decline and Fall of the Roman Empire* (Cambridge, Massachusetts, Harvard University Press, 1977), pp. 37–52, for insights into Gibbon's view of the precariousness of civilization. Gibbon's distinctive vision of the Middle Ages is explained in Arnaldo Momigliano, 'Gibbon from an Italian Point of View', in G. W. Bowersock, John Clive and Stephen R. Graubard (eds.), *Edward Gibbon and the Decline and Fall of the Roman*

Empire (Cambridge, Massachusetts, Harvard University Press, 1977), pp. 75–86.

Gibbon's views about time, history and progress should be set against the contemporary background of Enlightenment thought. Valuable interpretations are F. Manuel, *Shapes of Philosophical History* (London, Allen & Unwin, 1965); Henry Vyverberg, *Historical Pessimism in the French Enlightenment* (Cambridge, Massachusetts, Harvard University Press, 1958); Peter Burke, 'Tradition and Experience: The Ideas of Decline from Bruni to Gibbon', in G. W. Bowersock, John Clive and Stephen R. Graubard (eds.), *Edward Gibbon and the Decline and Fall of the Roman Empire* (Cambridge, Massachusetts, Harvard University Press, 1977), pp. 87–102.

Several scholars have assessed the *Decline and Fall* as a tract for the times, with an eye to the destiny of the British Empire. G. W. Bowersock, 'Gibbon on Civil War and Rebellion in the Decline of the Roman Empire', in G. W. Bowersock, John Clive and Stephen R. Graubard (eds.), *Edward Gibbon and the Decline and Fall of the Roman Empire* (Cambridge, Massachusetts, Harvard University Press, 1977), pp. 27–36 and Stephen R. Graubard, 'Edward Gibbon: Contraria Sunt Complementa', in G. W. Bowersock, John Clive and Stephen R. Graubard (eds.), *Edward Gibbon and the Decline and Fall of the Roman Empire* (Cambridge, Massachusetts, Harvard University Press, 1977), pp. 127–38, take a dim view of Gibbon's insights into the present. More sympathetic is L. P. Curtis, 'Gibbon's Paradise Lost', in F. W. Hiller (ed.), *The Age of Johnson: Essays Presented to C. B. Tinker* (New Haven, Yale University Press, 1949), pp. 73–90.

Conclusion: Making History

No systematic attempts have yet been made to examine the standing of Gibbon in the generations after his death and his place in the emergent history of English history-writing. A view of the development of nineteenth-century historiography is afforded in G. P. Gooch, *History and Historians in the Nineteenth Century* (London, Longmans, Green, 1913; 2nd edn, Longmans, Green, 1952), and F. Meinecke, *Historism*, trans. J. E. Anderson (London, Routledge & Kegan Paul, 1972).

Gibbon's 'philosophy of history' requires extensive exploration. For some introduction to recent debates about the nature of history see W. B. Gallie, *Philosophy and the Historical Understanding* (London, Chatto & Windus, 1964); Arthur C. Danto, *Analytical Philosophy of History* (Cambridge, Cambridge University Press, 1965); Morton White, *Foundations of Historical Knowledge* (London, Harper & Row,

1965); Maurice Mandelbaum, 'A Note on History as Narrative', *History and Theory*, 6 (1967), pp. 414–19; Patrick Gardiner (ed.), *Theories of History* (London, Allen & Unwin, 1959). Gibbon seems to have little to say to modern theorists of history. Perhaps that says more about them than about Gibbon. Why Gibbon appears now such a 'loner' is explored with great textual subtlety in W. B. Carnochan, *Gibbon's Solitude: The Inward World of the Historian* (Stanford, Stanford University Press, 1987).

The debate about why Rome declined and fell continues. For entries into the modern scholarship see Lynn White, Jr, 'The Temple of Jupiter Revisited', in Lynn White, Jr (ed.), *The Transformation of the Roman World* (Los Angeles, University of California Press, 1966), pp. 291–311; A. H. M. Jones, 'The Decline and Fall of the Roman Empire', *History*, 40 (1955), pp. 209–26; Perry Anderson, *Passages from Antiquity to Feudalism* (London, New Left Books, 1974) – a Marxist view; J. J. Saunders, 'The Debate on the Fall of Rome', *History*, 48 (1963), pp. 1–15; and Mortimer Chambers (ed.), *The Fall of Rome: Can It Be Explained?* (New York, Holt, Rinehart & Winston, 1963). Most recently see Peter Garnsey and Richard Saller, *The Roman Empire: Economy, Society and Culture* (Los Angeles, University of California Press, 1987).

Index

Abbreviation: EG = Edward Gibbon
All dates in parentheses are AD unless otherwise stated.

barbarians (*contd.*)
  Goths, 71, 82, 83–4, 85, 94, 116,
    136, 139, 140, 141, 142, 145,
    149, 150; Ostrogoths, 143;
    Visigoths, 84, 139
  Huns, 83, 84, 136, 139
  Merovingians, 84, 90, 143
  Monguls, 142
  Saracens, 85, 94, 130, 131
  Saxons, 84
  Sitones, 108–9
  Suevi, 139
  Suiones, 108–9
  Tartars, 149, 153
  Turks, 94, 145
  Vandals, 84, 124, 136, 139
barbarism, 135–57, 171, 172–3
Baridon, Michel, 165, 170
Barlaam, 15th-century Greek scholar,
  159
Baronio (Baronius), Cesare (1538–
  1607), 76, 78
  Annales Ecclesiastici (1588–1607),
    74
Barrell, John, 166
Bartélémy, Jean Jacques, Abbé (fl.
  1763), 35
Bayle, Pierre (1647–1706), 21, 46,
  50, 51, 68, 114, 156
Beausobre, Isaac de (fl. 1734), 76
Becker, Carl L., 6, 170
Beckford, William (d. 1799), 158
Beer, Sir Gavin de, 168
Behmen, Jacob, 112
Belisarius (505–65), Byzantine
  general, 105, 106
Bell, Robert H., 169
Bentham, Jeremy (1748–1832), 44
Bentley, Richard (1662–1742), 31,
  32, 54
Bernoulli, D. (1700–82), 156
Bible, 13, 20, 21–2, 23–4, 42, 45, 63,
  117, 120
Bibliothèque Raisonnée (1728–53), 58
Bibliothèque des Sciences, 57
Birch, Thomas (1705–66), 61
Bithynia, 96
Black, J. B., 4, 168
Blackstone, Sir William (1723–80),
  35
Blake, William (1757–1827), 137
Bletherie, Jean Philippe René de la
  (1696–1772), 35

Blount, Charles (1654–93), 20
Board of Trade, EG in, 101
Boccaccio (Boccace), Giovanni
  (1313–75), 159
Boethius (*c.* 480–524), 81
Bolingbroke, 1st Viscount, *see* Saint-
  John, Henry
Bond, Harold L., 88, 170
Bonnard, G. A., 165
Bontems, Madame Marie-Jeanne de
  Chatillon (1718–68), 58
Bossuet, Jacques Bénigne (1627–
  1704), 137
  Discourse on Universal History
    (1681), 25
  *Exposition of the Catholic
    Doctrine*, 114
  *History of the Protestant
    Variations*, 114
Boswell, James (1740–95), 65
  History of Corsica (1769), 63
Boulainvilliers, Henri de (1658–
  1722), 107, 109, 132
  Vie de Mahomed (1730), 130
Bouquet, Dom Martin (1685–1754),
  75
Bowdler, Thomas (1754–1825), 134
Bowersock, Glen W., 4, 167, 169,
  170, 171, 172, 173
Brady, Nicholas (1659–1726), 16
Braudel, Fernand, 6
Braudy, Leo, 170
Breitlinger, J. J. (fl. 1757), 49
Brislington House Lunatic Asylum, 1
Britain, Roman, independence of, 84
British constitution, 18, 27, 96, 153
British Empire, 161, 173
British Library, EG's unpublished
  papers, vii, 13, 165
British Museum, 32
Brontë, Charlotte (1816–55), 45
Brown, Peter, 137, 143, 172
Brownley, Martine Watson, 3, 166,
  171
Brumfitt, J. H., 170
Buck, Willis R., 9–10, 166
Burke, Edmund (1729–97), 26, 34,
  109, 154
Burke, Peter, 173
Burney, Charles (1726–1814), 30
Burney, Frances, *see* Arblay, Madame
  d'
Burrow, J. W., 171

INDEX